# Preface

The aim of producing this text was to provide a brief introduction to the theory, practice, and applications of X-ray photoelectron spectroscopy and Auger electron spectroscopy. These techniques of surface analysis have now been available for almost twenty years, during which time they have become accepted methods for the analysis of solid surfaces, and have now reached a state of maturity. With this status has come the need to provide an introductory text giving the bare bones of the technique to potential users, from which the principles of both the atomic processes and the operation of the hardware can be grasped. This has been the intention in the preparation of the major part of this handbook. In addition, most scientists delving into an analytical method, which is new to them, will wish to explore the possibilities of applying it in their particular field. It is with this in mind that the chapter on applications has been written. The use of XPS and AES to yield electronic information of interest to the solid state chemist or physicist has been deliberately excluded and the emphasis is unashamedly on materials science, with examples drawn from those areas in which electron spectroscopy has made the greatest impact.

The number of references has been kept to a minimum, and, in contrast with the traditional method of citation, are grouped together as the bibliography. One or two key references, in each subject area, are provided as a springboard for further reading.

*University of Surrey*                                                                 J.F.W.
January 1990

# An Introduction to Surface Analysis by Electron Spectroscopy

John F. Watts

Department of Materials Science and Engineering,
University of Surrey,
Guildford, UK

Oxford University Press · Royal Microscopical Society · 1990

Oxford University Press, Walton Street, Oxford OX2 6DP

Oxford New York Toronto
Delhi Bombay Calcutta Madras Karachi
Petaling Jaya Singapore Hong Kong Tokyo
Nairobi Dar es Salaam Cape Town
Melbourne Auckland

and associated companies in
Berlin Ibadan

Royal Microscopical Society
37/38 St Clements
Oxford OX4 1AJ

Oxford is a trade mark of Oxford University Press

Published in the United States
by Oxford University Press, New York

British Library Cataloguing in Publication Data
Watts, John F.
An introduction to surface analysis by electron
spectroscopy.
1. Solids. Surfaces. Electron spectroscopy
I. Title   II. Series
530.41
ISBN 0-19-856425-2

Library of Congress Cataloging in Publication Data
Watts, John F.
An introduction to surface analysis by electron spectroscopy/
John F. Watts.
p.      cm. — (Microscopy handbooks: 22)
Includes bibliographical references.
1. Surfaces (Technology)—Analysis.   2. Electron spectroscopy.
I. Title.   II. Series.
TP156.S95W37 1990      620'.44—dc20      90-35779
ISBN 0-19-856425-2

Printed in Great Britain by
Bookcraft (Bath) Ltd
Midsomer Norton, Avon

# Acknowledgements

It is a pleasure to thank all the staff and students of The Surface Analysis Laboratory at the University of Surrey for providing such a stimulating working environment over the years. Many people have provided (knowingly or unknowingly!) ideas, information, and results that are included in the following pages. My particular thanks go to Professor Jim Castle, Drs Mark Baker, Pam Budd, Judith Cohen, Steve Greaves, Ke Ruoru, Peter Mills, Sally Mugford, and Roy Paynter. I offer my apologies if I have not done them justice.

Certain figures are reproduced with permission of the copyright holders as detailed below;
Fig. 3.2a  Academic Press Inc.
Fig. 3.2b  The Royal Microscopical Society
Fig. 4.1  Elsevier Applied Science Publishers Ltd
Fig. 4.2  Chapman and Hall Ltd
Fig. 4.3  The Royal Microscopical Society
Fig. 5.4  The Institute of Metals
Fig. 5.6  John Wiley and Sons Ltd
Fig. 5.7  John Wiley and Sons Ltd
Fig. 5.8  Academic Press Inc.
Fig. 5.10  John Wiley and Sons Ltd
Appendix 1  Chart of principal Auger electron energies. Reproduced from *Handbook of Auger electron spectroscopy* (1978) (ed. L. E. Davis, N. C. MacDonald, P. W. Palmberg, G. E. Riach, and R. E. Weber). Published by Physical Electronics Division, Perkin-Elmer Corporation, 6509 Flying Cloud Road, Eden Prairie, MN 55343, USA.

# Contents

Appendix 1  Chart of principal Auger electron energies

Appendix 2  Table of binding energies for XPS accessible with
AlK$\alpha$ radiation

# 1 Electron spectroscopy: some basic concepts

In electron spectroscopy we are concerned with the emission and energy analysis of low energy electrons (generally in the range 20–2000 eV).* These electrons are liberated from the specimen being examined as a result of the photoemission process (in X-ray photoelectron spectroscopy, XPS) or the radiationless de-excitation of an ionized atom by the Auger emission process (in Auger electron spectroscopy (AES), and scanning Auger microscopy (SAM)). In the simplest terms, an electron spectrometer consists of the sample under investigation, a source of primary radiation, and an electron analyser all contained within a vacuum chamber preferably operating in the ultra-high vacuum regime. In practice, there will usually be a secondary chamber fitted with various sample preparation facilities and perhaps ancillary analytical facilities. A datasystem will often be used for the acquisition and subsequent processing of data. The source of the primary radiation for the two methods is different; X-ray photoelectron spectroscopy making use of soft X-rays, generally AlKα or MgKα, whereas AES and SAM rely on the use of an electron gun. The specification for electron guns used in Auger analysis varies tremendously, particularly as far as the spatial resolution is concerned, which for finely focused guns may be between 5 $\mu$m and 15 nm. In principle, the same energy analyser may be used for both XPS and AES; consequently, the two techniques are often to be found in the same analytical instrument. Before considering the uses and applications of the two methods, a brief review of the basic physics of the two processes and the strengths and weaknesses of each technique will be given.

## 1.1 X-ray photoelectron spectroscopy (XPS)

In XPS we are concerned with a special form of photoemission i.e. the ejection of an electron from a core level by an X-ray photon of energy $h\nu$. The energy of the emitted photoelectrons is then analysed by the electron spectrometer and the data presented as a graph of intensity (or counts per second) versus electron energy; the X-ray induced photoelectron spectrum.

The kinetic energy ($E_K$) of the electron is the experimental quantity measured by the spectrometer, but this is dependent on the energy of the

---

*Units: In electron spectroscopy, energies are expressed in the non-SI unit the electron volt. The conversion factor to the appropriate SI unit is 1 eV = $1.595 \times 10^{-19}$ J.

X-ray source employed and is therefore not an intrinsic material property. The binding energy of the electron ($E_B$) is the parameter which identifies the electron specifically, both in terms of its parent element and atomic energy level. The relationship between the parameters involved in the XPS experiment is as follows:

$$E_B = h\nu - E_K - W$$

where h$\nu$ is the photon energy, $E_K$ is the kinetic energy of the electron, and $W$ is the spectrometer work function.

As all three quantities on the right-hand side of the equation are known or measurable, it is a simple matter to calculate the binding energy of the electron. In practice, this task will be performed by the control electronics associated with the spectrometer and the operator merely selects a binding or kinetic energy scale whichever is considered the more appropriate.

The process of photoemission is shown schematically in Fig. 1.1, where an electron from the K shell is ejected from the atom; this electron is termed a 1s photoelectron. The photoelectron spectrum will reproduce the electronic structure of an element quite accurately as all electrons with a binding energy less than the photon energy will feature in the spectrum. This is illustrated in Fig. 1.2 where the XPS spectrum of gold is superimposed on a representation of the electron orbitals. Those electrons which are excited and escape without energy loss contribute to the characteristic peaks in the spectrum; those which undergo inelastic scattering and suffer energy loss contribute to the *background* of the spectrum. Once a photoelectron has been emitted, the ionized atom must relax in some way. This can be achieved by the emission of an X-ray photon, known as X-ray fluorescence. The other possibility is the ejection of an electron as an Auger electron. Thus Auger electrons are produced as a consequence of the XPS process, often referred to as X-AES (X-ray induced Auger electron spectroscopy). X-AES, although not widely

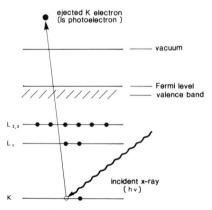

Fig. 1.1. Schematic of the XPS process, showing photoionization of an atom by the ejection of a 1s electron.

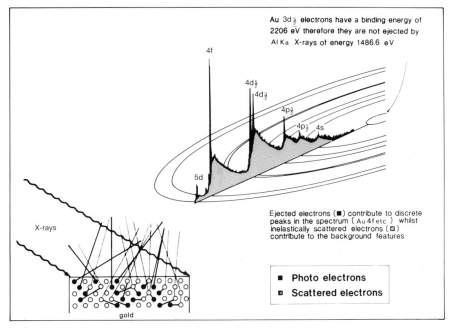

Fig. 1.2. Photo-electron spectrum of gold showing the manner in which electrons escaping from the solid can contribute to the discrete peaks or suffer energy loss and contribute to the background. The spectrum is superimposed on a schematic of the electronic structure of gold to illustrate how each orbital gives rise to photoelectron lines.

practised, can yield valuable chemical information about an atom. For the time being we will restrict our thoughts to AES in its more common form, that is when the Auger electrons are excited by a finely focused electron beam.

The nomenclature employed to describe photoelectrons and Auger electrons is different, XPS uses the so-called spectroscopists' or chemists' notation in which electrons are described by the appropriate quantum numbers, i.e. 1s, $2p_{3/2}$, $3d_{5/2}$, $4f_{7/2}$, whilst Auger electrons are identified by the X-ray notation (shell name) i.e. $K$, $L$, $M$, $N$, and so on.

## 1.2 Auger electron spectroscopy (AES)

When a specimen is irradiated with electrons, core electrons are ejected in the same way that an X-ray beam will cause core electrons to be ejected in XPS. The difference is that in the case of electron irradiation the secondary electrons contain no analytical information—although those of low energy are very useful for imaging purposes as in scanning electron microscopy. However, once an atom has been ionized it must in some way return to its ground state. The emission of an X-ray photon may also occur, which is the basis of electron probe microanalysis (EPMA), carried out in many electron microscopes by either energy or wavelength dispersive spectrometers. The other possibility is that the core hole (for instance a $K$ shell vacancy as shown in Fig. 1.1) may be filled by an electron from a higher level, the $L_{2,3}$ level in Fig. 1.3. In order to conform with the principle of the conservation of energy, another electron must be ejected from the atom, e.g. another $L_{2,3}$ electron in the schematic of Fig. 1.3. This electron is termed the $KL_{2,3}L_{2,3}$ Auger electron. Its kinetic energy is approximately equal to the difference between the energy level of the core hole and the energy levels of the two outer electrons ($E_{L_{2,3}}$):

$$E_{KL_{2,3}L_{2,3}} \approx E_K - E_{L_{2,3}} - E_{L_{2,3}}$$

This equation does not take into account the interaction energies between the core holes ($L_{2,3}$ and $L_{2,3}$) in the final atomic state nor the inter- and extra-relaxation energies which come about as a result of the additional core screening needed. Clearly, the calculation of the energy of Auger electron transitions is much more complex than the simple model outlined above, but there is a satisfactory empirical approach which considers the energies of the atomic levels involved and those of the next element in the periodic table.

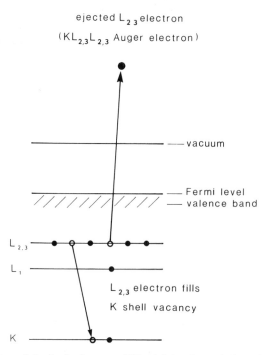

Fig. 1.3. Relaxation of the ionized atom of Fig. 1.1 by the emission of a $KL_{2,3}L_{2,3}$ Auger electron.

Following this empirical approach, the Auger electron energy of transition $KL_1L_{2,3}$ for an atom of atomic number $Z$ is written:

$$E_{KL_1L_{2,3}}(Z) = E_K(Z) - \tfrac{1}{2}[E_{L_1}(Z) + E_{L_1}(Z+1)] - \tfrac{1}{2}[E_{L_{2,3}}(Z) + E_{L_{2,3}}(Z+1)]$$

Clearly for the $KL_{2,3}L_{2,3}$ transition the second and third terms of the above equation are identical and the expression is simplified to:

$$E_{KL_{2,3}L_{2,3}}(Z) = E_K(Z) - [E_{L_{2,3}}(Z) + E_{L_{2,3}}(Z+1)]$$

It is the kinetic energy of this Auger electron ($E_{KL_{2,3}L_{2,3}}$) that is the characteristic material quantity irrespective of the primary beam composition (i.e. electrons, X-rays, ions) or its energy. For this reason Auger spectra are always plotted on a kinetic energy scale. The use of a finely focused electron beam for AES enables us to achieve surface analysis at a high spatial resolution, in a manner analogous to EPMA in the scanning electron microscope. By combining an electron spectrometer with an ultra-high vacuum (UHV) SEM it becomes possible to carry out scanning Auger microscopy. In this mode of operation various imaging and chemical mapping procedures become possible.

## 1.3 Scanning Auger microscopy (SAM)

In the scanning Auger microscope various modes of operation are available; the variable quantities being the position of the electron probe on the specimen ($x$ and $y$) and the setting of the electron energy analyser ($E$) corresponding to the energy of emitted electrons to be analysed. The various possibilities are summarized in Table 1.1.

Table 1.1. *Modes of analysis available with SAM*

| Mode of analysis | Scanned | Fixed |
|---|---|---|
| Point analysis | $E$ | $x,y$ |
| Line scan | $x$ | $E,y$ |
| Chemical map | $x,y$ | $E$ |

A further possibility not covered in Table 1.1 is the control of both spectrometer and electron beam position by a microprocessor. The provision of datasystems for electron spectroscopy is desirable and is covered in the next chapter.

As the Auger electron yield is very sensitive to electron take-off angle, an image of Auger electron intensities will invariably reflect the surface topography of the specimen, possibly more strongly than the chemical variations, as illustrated (Fig. 1.4) in the Auger map of carbon fibres (Fig. 1.4(b)) which is very similar to the SEM image (Fig. 1.4(a)). The problem is

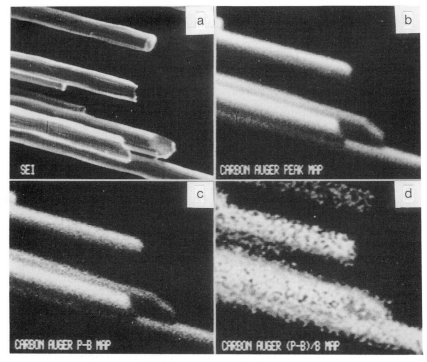

Fig. 1.4. Scanning Auger microscopy of carbon fibres. (a) SEM image, (b) peak map (P) of carbon Auger electrons, (c) peak–background map (P–B), B recorded 40 eV from Auger peak, (d) correction for topographic effects using (P–B)/B algorithm. The diameter of the fibres is 7 $\mu$m.

overcome by recording a background (B) as well as the Auger peak (P) map. However, a simple subtraction of the background counts from the peak intensity (P–B) is not sufficient as shown by the (P–B) map of Fig. 1.4(c). The use of a simple algorithm such as (P–B)/B, allows correction for the effects of surface topography, Fig. 1.4(d), where variation in intensity due to the cylindrical shape of the fibres has been completely suppressed and only chemical information remains.

## 1.4 The depth of analysis in electron spectroscopy

The depth of analysis in both XPS and AES varies with the kinetic energy of the electrons under consideration. The inelastic mean free path ($\lambda$) of an electron varies as $E^{0.5}$ in the energy range of interest in electron spectroscopy and various relationships have been suggested which relate $\lambda$ to electron energy and material properties. One such equation proposed by Seah and Dench (1979) of the National Physical Laboratory is given below:

$$\lambda = \frac{538 a_A}{E_A^2} + 0.41 a_A (a_A E_A)^{0.5}$$

where $E_A$ is the energy of the electron in eV, $a_A^3$ is the volume of the atom in nm³, $\lambda$ is in nm. The intensity of electrons ($I$) emitted from a depth ($d$) is given by the Beer–Lambert relationship:

$$I = I_0 \exp(-d/\lambda \sin \theta)$$

where $I_0$ is the intensity from an infinitely thick clean substrate, $\theta$ is the electron take-off angle relative to the sample surface. The variation of electron intensity with depth is shown schematically, for a carbon substrate, in Fig. 1.5.

The Beer–Lambert equation can be manipulated in a variety of ways to provide information about overlayer thickness and to provide a non-destructive depth profile (that is without removing material by mechanical, chemical, or ion-milling methods). Using the appropriate analysis of the above equation, it can be shown that by considering electrons that emerge at 90° to the sample surface, some 65 per cent of the signal in electron spectroscopy will emanate from a depth of $\lambda$, 85 per cent from a depth of $2\lambda$, and 95 per cent from a depth of $3\lambda$, as illustrated in Fig. 1.5. As values of the inelastic mean free paths are of the order of a few nanometres the surface sensitivity of XPS and AES can easily be appreciated. We shall return to the question of the development of compositional depth profiles by making use of the Beer–Lambert expression in Chapter 3.

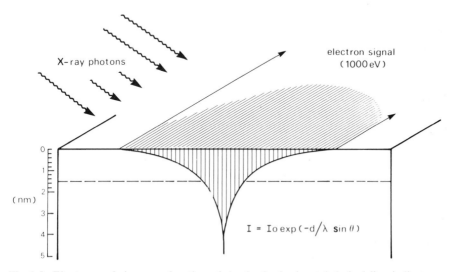

Fig. 1.5. Electron emission as a function of depth, the horizontal dashed line indicates a distance from the surface of the inelastic mean free path ($\lambda$).

## 1.5  Comparison of XPS and AES/SAM

Although it is difficult to make a comparison of techniques before they are described and discussed in detail, it is pertinent at this point to outline the strengths and weaknesses of each to provide background information.

XPS is also known by the acronym ESCA (electron spectroscopy for chemical analysis). It is this *chemical* specificity which is the major strength of XPS as an analytical technique, one for which it has become deservedly popular. By this we mean the ability to identify not only the elements present in the analysis but also the chemical state. In the case of iron, for instance, the spectra of $Fe^0$, $Fe^{2+}$, and $Fe^{3+}$ are all slightly different and to the expert eye are easily distinguishable. However, such information is attainable in XPS only at the expense of spatial resolution, and XPS is usually regarded as an area averaging technique with some 10 $mm^2$ contributing to the analysis. Small area XPS (SAXPS or even SAX) is available on most modern instruments and when operating in this mode a spatial resolution of 150 $\mu$m is possible. Set beside a spatial resolution of 15–20 nm which can be achieved on the latest commercial Auger microprobes, it becomes clear that XPS is not the way to proceed for surface analysis at high spatial resolution, but the advantages of the levels of information available from an XPS analysis (the ease with which a quantitative analysis can be achieved, its applicability to insulators, and the ready availability of chemical state information) will often offset this. In addition to the chemical state information referred to above, XPS spectra can be quantified in a very straightforward manner and meaningful comparisons can be made between specimens of a similar type. Quantification of Auger data is rather more complex and the accuracy obtained is generally not as good. Because of the complementary nature of the two methods and the ease with which Auger and photoelectron analyses can be made on the same instrument, the two methods have come to be regarded as the most important methods of surface analysis in the context of materials science. All manufacturers of electron spectrometers offer both XPS and AES options for their systems.

## 1.6  The availability of surface analytical equipment

The capital cost of an XPS/AES/SAM spectrometer is high when compared with most electron microscopes, and is of the order of £0.5 M for a comprehensive system. This, allied to the fairly steep learning curve that the newcomer must ascend before confidence in the technique is obtained, has lead to the development of laboratories offering surface analysis as a service facility, both in Western Europe and the USA. In the UK, they are generally associated with universities or polytechnics but the balance between academic and industrial work varies greatly. The use of service facilities presents a very attractive proposition to inexperienced users in that expert advice is always on hand to ensure the efficient use of instrument time; a factor that is of paramount importance as the daily charge for the use of such a facility can sometimes exceed £1000. It is not unusual for analysts to 'cut their teeth' on the field of surface analysis in such a way; once the need within their company (and their own personal expertise) has been established, a surface analysis system can be specified for their own particular needs.

Although originally the exclusive preserve of research laboratories and academic institutions, surface analysis facilities are now frequently to be found in trouble shooting and quality assurance roles. As the techniques find wider applications, so the market grows and manufacturers are only too willing to continue developing their spectrometers. Thus the future for XPS and AES seems assured well into the next century.

# 2 Electron spectrometer design

The design and construction of electron spectrometers is a very complex undertaking and will usually be left to one of the handful of specialist manufacturers world-wide although many users will specify minor modifications to suit their own requirements. The various modules necessary for analysis by electron spectroscopy are (in addition to a specimen): a source of the primary beam (either X-rays or electrons); an electron energy analyser and detection system, all contained within a vacuum chamber; and a data-system which is nowadays considered as an integral part of the system.

## 2.1 The vacuum system

All commercial spectrometers are now based on vacuum systems designed to operate in the ultra-high vacuum (UHV) range of $10^{-8}$ to $10^{-10}$ mbar, and it is now generally accepted that XPS and AES experiments must be carried out in this pressure range. The reason for this is twofold; first, the low energy electrons are easily scattered by the residual gas molecules and unless their concentration is kept to an acceptable level the total spectral intensity will decrease, whilst the noise present within the spectrum will increase. The second, and perhaps more important reason, results from the very surface sensitivity of the techniques themselves. Even at $10^{-6}$ mbar, it is possible for a monolayer of gas to be adsorbed onto a solid surface in about one second; setting this against a typical acquisition time of around 900 seconds and a detection limit of about 1 per cent monolayer, that is 1 in 100 of the atoms in the surface atomic layer, clearly establishes the need for a UHV environment during analysis. The manner in which such vacua are established will depend on customer and manufacturer preferences. The chambers and associated piping will invariably be made of stainless steel, and joints will usually be effected using crushed copper gaskets (a system generally referred to as *conflat* following the Varian Associates designation who own the trademark). The trajectory of the electrons is strongly influenced by the Earth's magnetic field, consequently some form of magnetic screening is required around the sample and electron analyser. There are two approaches to this problem. The most elegant solution is to fabricate the entire analysis chamber from a material with low magnetic permeability ($\mu$-metal); an acceptable alternative is to fabricate shielding panels, either as sleeving within the instrument or as a bolt-on outer shroud. The methodology depends on the manufacturer. In addition, compensation coils may be arranged around the analyser and transfer lens to mitigate the effect of such magnetic fields. The most popular way of achieving UHV conditions is by means of a diffusion pump charged with polyphenylether oil and fitted with a liquid nitrogen cooled cold trap. Other alternatives include ion-pumps, turbomolecular pumps, cryopumps, and as an auxiliary system, titanium sublimation pumps. All UHV systems need baking from time to time to remove adsorbed layers from the chamber walls, the baking temperature is dictated by the analytical options fitted to the spectrometer but is usually in the range 100–160°C for routine use.

## 2.2  The sample

Although electron spectroscopy can be carried out successfully on gases and liquids as well as solids, gases and liquids necessarily yield bulk molecular and chemical information rather than surface chemical information. Consequently, we shall restrict our discussions here to solid samples. The criteria for analyses by AES and XPS are not the same; the requirements for specimens for Auger spectroscopy being somewhat more stringent. Samples for both XPS and AES must be stable within the UHV chamber of the spectrometer, the two types of specimen which can pose problems here are very porous materials (such as some ceramic and polymeric materials) and those with low vapour pressure which volatilize easily (for example some organic compounds). As far as XPS is concerned once these requirements have been fulfilled the sample is amenable to analysis. For Auger analysis, however, the use of an electron beam dictates that for routine analysis the specimen should be conducting and effectively earthed in addition to the vacuum compatability requirements outlined above. As a guide, if a specimen can be imaged (in an uncoated condition) in an SEM without any charging problems, a specimen of a similar type can be analysed by Auger electron spectroscopy. The analysis of insulators such as polymers and ceramics by AES is quite feasible but its success relies heavily on the skill and experience of the instrument operator. Such analysis is achieved by ensuring that the incoming beam current is exactly balanced by the combined current of emitted electrons (all secondaries including Auger electrons, backscattered and elastically scattered electrons etc.), by optimizing beam energy (3–5 kV), specimen current (very low probably <10 nA), and electron take off angle.

The mounting of conducting specimens is best achieved with clips or bolt down assemblies although for XPS the use of double sided adhesive tape has much to recommend it. For conducting specimens, a fine strip of conducting paint, in addition to the adhesive tape, is all that is necessary to prevent sample charging. Alternatively, metal tape with a metal loaded (conducting) adhesive may be used. Most laboratories have a selection of sample holders, usually fabricated in house, to accommodate large and awkwardly shaped specimens. Discontiguous specimens present rather special problems. In the case of powders, the best method is embedding in indium foil, but if this is not feasible, dusting onto double sided adhesive tape can be a very satisfactory alternative. Fibres and ribbons can be mounted across a gap in a specimen holder ensuring that no signal from the mount is detected in the analysis.

The type of specimen mount varies with instrument design and most modern spectrometers use a sample stub similar to the type employed in scanning electron microscopy. For analysis, the sample is held in a high resolution manipulator with $x$, $y$, and $z$ translations, and rotation, $\theta$, about the $z$ axis. For scanning Auger microscopy where the time taken to acquire high resolution maps is perhaps an hour or so, the stability of the stage is critical, since any drift during analysis will degrade the resolution of the images. For angular resolved XPS, the amount of backlash in the rotary drive must be small and the scale should be graduated in increments of two degrees.

Once mounted for analysis, heating or cooling of the specimen can be carried out *in vacuo*. Cooling is generally restricted to liquid nitrogen temperatures although liquid helium stages are available. Heating may be achieved by direct (contact) heating using a small resistance heater or by electron bombardment for higher temperatures. Such heating and cooling will either be a preliminary to analysis or carried out during the analysis itself (with the obvious exception of electron bombardment heating). Heating in particular will often be carried out in a preparation chamber because of the possibility of severe outgassing encountered at higher temperatures.

The routine analysis of specimens by AES and in particular XPS is a very time consuming business and some form of automation is clearly desirable. This is available from several manufacturers in the form of a computer driven carousel or table which enables a batch of specimens to be analysed when the machine is left unattended, typically overnight.

A modern commercial electron spectrometer is illustrated in Fig. 2.1 and a schematic diagram of the analysis chamber in Fig. 2.2.

## 2.3 The X-ray source for XPS

The choice of anode material for XPS is determined by the energy of the X-ray transition generated. It must be of high enough photon energy to excite an intense photoelectron peak from all elements of the periodic table (with the exception of the very lightest); it must also possess a natural X-ray line width that will not broaden the resultant spectrum excessively. The most popular anode materials are aluminium and magnesium; they are available in a single X-ray gun with twin anode configuration which provides AlK$\alpha$ and MgK$\alpha$ photons of energy 1486.6 eV and 1253.6 eV respectively. Such twin anode assemblies have a separate filament for each anode and changing from

Fig. 2.1. A modern electron spectrometer.

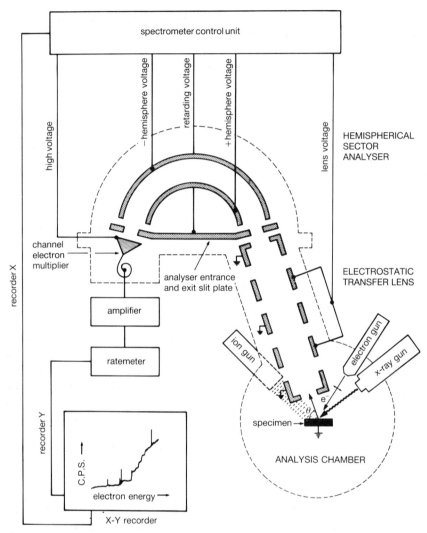

Fig. 2.2. Schematic of the analysis chamber and analyser of the spectrometer of Fig. 2.1.

one source to the other is accomplished by the simple expedient of switching power to the appropriate filament. This is possible because, unlike X-ray diffraction anodes, it is the anode and not the filament which is at a high potential (for XRD the filament is at a high negative potential and the anode at ground; for XPS the filament is at or near ground and the anode at a high positive potential of 10–15 kV). Such twin anode assemblies are useful as they provide a modest depth profiling capability—the difference in the analysis depth in organic materials of the carbon 1s electron is about 1 nm greater for electrons excited by AlKα, (analysis depth approximately

7 $\cos\theta$ nm) compared with MgK$\alpha$ (analysis depth approximately 6 $\cos\theta$ nm)—and provide the ability to differentiate between Auger and photoelectron transitions when the two overlap in one radiation. XPS peaks will change to a position 233 eV higher on a kinetic energy scale on switching from MgK$\alpha$ to AlK$\alpha$ whereas the energy of Auger transitions remains constant.

Several other materials have been used successfully as X-ray anodes, most notably SiK$\alpha$ (h$\nu$ = 1739.5 eV), ZrL$\alpha$ (h$\nu$ = 2042.4 eV), AgL$\alpha$ (h$\nu$ = 2984.3 eV) and TiK$\alpha$ (h$\nu$ = 4510.0 eV), all of which can be combined with aluminium or magnesium in the standard twin anode configuration. There are two advantages of these higher energy anodes. First, energy levels not available in conventional XPS become accessible—in AlK$\alpha$ radiation the Mg 1s electron is the highest K electron attainable, in SiK$\alpha$ this is extended to the Al 1s electron, in ZrL$\alpha$ the Si 1s electron, in AgL$\alpha$ the Cl 1s electron and in TiK$\alpha$ the Ca 1s electron. Because the use of higher energy photon sources increases the kinetic energy of the ejected photoelectrons available when compared with conventional XPS, higher energy XPS provides a non-destructive means of increasing the analysis depth. It is therefore possible to build up a depth profile of a specimen merely by changing the X-ray source and monitoring the apparent change in composition.

A means of reducing the natural line width is by monochromation. This is achieved by the dispersion of the X-rays by a suitable crystal. At present, the only commercially available X-ray monochromators used for XPS employ a natural quartz crystal. This is only suitable for AlK$\alpha$ radiation (first order reflections) or AgL$\alpha$ (second order reflections). The primary reason for using monochromated radiation is the reduction in X-ray line width, from 0.85 eV to approximately 0.4 eV for AlK$\alpha$, and from 2.6 eV to 1.2 eV for AgL$\alpha$. However, there is an added bonus in that the unwanted portions of the X-ray spectrum i.e. the Bremsstrahlung continuum and satellite peaks, are also removed. As AlK$\alpha$ and AgL$\alpha$ use the same monochromator assembly they are available together in a twin anode assembly. In monochromated XPS, the X-ray flux falling on the specimen is greatly reduced so the acquisition time for spectra of an acceptable quality must be increased and under certain conditions such as low elemental concentrations it is preferable to use an achromatic source followed by deconvolution using a computer routine.

Monochromation effectively reduces the X-ray 'footprint' on the specimen to a line source typically 6 mm $\times$ 1 mm. This can be reduced still further by using a finely focused electron beam to excite the X-rays at the anode. By the use of an X-ray monochromator of this type it is possible to produce a small area XPS (SAX) analysis and this now forms the basis of a commercial instrument with a spatial resolution of 150 $\mu$m. This is one of several routes to SAX, the other commercially available method, electron-optical aperturing, is discussed later in this chapter (p. 22).

## 2.4 The electron gun for AES

Since 1969, (the year in which Auger electron spectrometers became commercially available) AES has become a well accepted analytical method for the provision of surface analyses at high spatial resolution and the intervening years have seen the electron guns improve from the 500 $\mu$m resolution of the converted oscilloscope gun of the late 1960s to the 15 nm resolution which has been demonstrated on a field emission gun for AES in the early 1990s. In between these two extremes are the 5 $\mu$m and 200 nm guns which form the mainstay of most Auger systems in use today.

The critical components of the electron gun are the electron source and the lens assemblies for beam focusing, shaping, and scanning. Electron sources may be either thermionic emitters or field emitters, lenses for the electron gun may be electrostatic or, for high resolution applications, electromagnetic. If we consider the lenses first, the criterion which distinguishes an electron gun for AES from one for conventional electron microscopy is the need to operate in an UHV environment. In the early days of AES, this effectively precluded the use of electromagnetic lenses as the coils were not able to withstand UHV bakeout temperatures. Consequently, electrostatic lenses were much favoured and by gradual design improvements the stage has now been reached where electron guns with electrostatic lenses can achieve a spatial resolution of 200 nm. The major step forward as far as scanning Auger microscopy was concerned, was the development of a gun with bakeable electromagnetic lenses; by the late 1970s, such lenses, with a spatial resolution of 50 nm and a thermionic emitter were routinely available for the first time.

For Auger electron spectroscopy it is clearly essential that the electron flux is stable over relatively long periods of time and it was for this reason that until relatively recently the designers of electron guns for surface analysis were reluctant to use field emission guns. Electron guns with low to medium spatial resolution ($\geq 200$ nm) will invariably feature a thermionic emitter. The simplest form of thermionic source is a tungsten wire fabricated in the form of a hairpin. When an electric current is passed through it, the temperature rises giving electrons sufficient energy to overcome the work function and be released into free space. The work function is the energy required for an electron to escape from a solid surface; for tungsten this is about 4.5 eV. By reducing the work function, we can increase the number of electrons emitted per unit area per unit solid angle, thereby increasing the so-called *brightness* of the source. The most widely used material for high-brightness sources is single crystal lanthanum hexaboride ($LaB_6$) and filaments made from this material are a valuable addition to electron guns with optimized electrostatic or electromagnetic lenses.

The other type of gun which finds occasional use in Auger microscopy is that based on the field emission source. The operating principle of a field emission source is not to give the electrons sufficient energy to jump the work function barrier (as in the thermionic process) but to reduce the magnitude of the barrier itself, both in height, (marginally), but more importantly in width. It is the latter factor which leads to improved electron emission, as electrons from the Fermi level can penetrate the barrier by quantum mechanical tunnelling and thus escape from the emitter with no loss in energy. Another characteristic of the field emission source is its narrow energy distribution, as there are no electrons above the Fermi level and those below it have a rapidly decreasing probability of escape. In practical terms, this feature means a smaller spot size as chromatic aberrations within the electron lenses are reduced. Usually, field emission is achieved by the application of a very strong electrostatic field between the filament, which itself must be in the form of a needle with a tip radius of approximately 50 nm, and an extraction electrode. The small area of emission from the tip into a small solid angle provides a high brightness compared with thermionic sources, although the total current will be somewhat lower. The filament material employed is a tungsten single crystal and cleanliness is essential; this is maintained by using differential pumping at the electron source and by periodically flashing the crystal to remove adsorbed gases. One of the main disadvantages of single crystal filaments, be they thermionic or field emitters, is the need to keep them scrupulously clean. This is one of the factors which has ensured the continued popularity of the tungsten hairpin as an electron source where the requirements are less stringent.

The spot size attainable with a particular electron gun is a function of the primary beam current. For example the smallest spot size obtainable on a scanning Auger microscope with electromagnetic lenses and $LaB_6$ filament is 20 nm at 0.1 nA but increases to 100 nm at 10 nA. The intensity of Auger electrons emitted depends on the specimen current and at 0.1 nA spectrum acquisition will be a lengthy process, but at 10 nA the current of Auger electrons will be greatly increased. A satisfactory compromise must be reached between spatial resolution and spectral intensity. A field emission source in a similar column will give somewhat better resolution but at a much improved current owing to superior brightness. Such electron guns are the preserve of high resolution scanning Auger microscopes. For routine SAM, the best configuration is probably a high brightness 200 nm gun with $LaB_6$ source and electrostatic lenses of the type illustrated in Fig. 2.3. Such an assembly will give reliable routine operation with the minimum of attention.

## 2.5 Analysers for electron spectroscopy

There are two types of electron energy analyser in general use for XPS and AES, the cylindrical mirror analyser (CMA) and the hemispherical sector analyser (HSA). These two types have been developed independently and

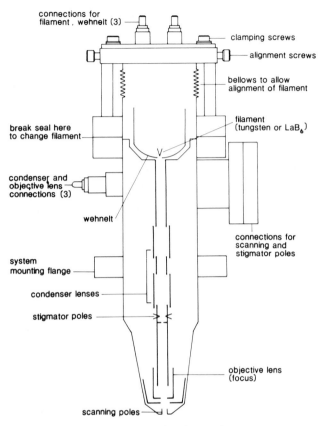

connections for
filament , wehnelt (3)

clamping screws

alignment screws

bellows to allow
alignment of filament

break seal here
to change filament

filament
(tungsten or LaB$_6$)

condenser and
objective lens
connections (3)

wehnelt

connections for
scanning and
stigmator poles

system
mounting flange

condenser lenses

stigmator poles

objective lens
(focus)

scanning poles

Fig. 2.3. Schematic of a UHV electron gun for Auger electron spectroscopy. The system mounting flange has a diameter of 114 mm.

until quite recently the CMA was preferred for AES and the HSA for XPS. The development of these two analysers is a reflection of the requirements of AES and XPS at the inception of the techniques. The primary requirement for Auger spectroscopy was that of high sensitivity (analyser transmission); the intrinsic resolution (the contribution of analyser broadening to the resultant spectrum) being of minor importance. The need for high sensitivity led to the development of the CMA. For XPS, on the other hand, it is spectral resolution that is the cornerstone of the technique and this led to the development of the HSA as a design of analyser with sufficiently good resolution. None the less, very high quality XPS work can be obtained from a suitably modified CMA Auger system and similarly good Auger spectra can be obtained from HSA photoelectron spectroscopy installations. However, with the advent of high spatial and high spectral resolution Auger electron spectroscopy the popularity of the CMA is declining for reasons discussed later.

## The cylindrical mirror analyser

Although the early Auger spectra were recorded using retarding field analysers of the type employed for low energy electron diffraction (LEED) it was the development of the CMA which made a major contribution to the popularity of AES as an analytical method.

The CMA basically consists of two concentric cylinders as illustrated in Fig. 2.4, the inner cylinder is held at earth whilst the outer is ramped at a negative potential. The usual practice is to mount the electron gun co-axially within the analyser as illustrated in Fig. 2.4. A certain proportion of the Auger electrons emitted will pass through the defining aperture in the inner cylinder, and, depending on the potential applied to the outer cylinder, electrons of the desired energy will pass through the detector aperture and be re-focused at the electron detector (a channel electron multiplier). Thus an energy spectrum—the direct energy spectrum—can be built up by merely scanning the potential on the outer cylinder to produce a spectrum of intensity (in counts per second) versus electron kinetic energy. This spectrum will contain not only Auger electrons but all the other emitted electrons, the Auger peaks being superimposed, as weak features, on an

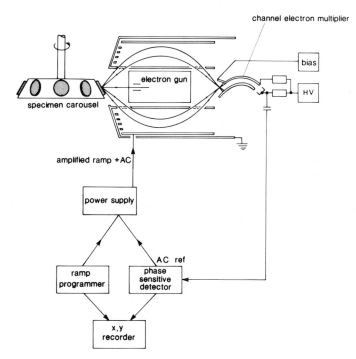

Fig. 2.4. An Auger electron spectroscopy system based on a single pass CMA electron energy analyser.

intense background. For this reason it has become common practice to record the differential spectrum rather than the direct energy spectrum. This is achieved by applying a small a.c. modulation to the analyser and comparing the output at the channeltron with this standard a.c. signal by means of a phase sensitive detector (lock-in-amplifier). The resultant signal is displayed as the differential spectrum. A comparison of direct (pulse counted) and differential Auger electron spectra, from a copper foil, is presented in Fig. 2.5.

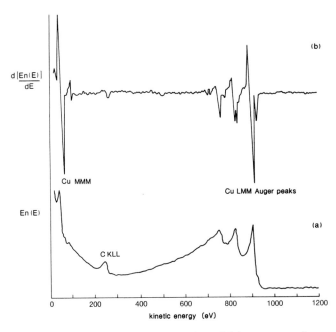

Fig. 2.5. Comparison of (a) direct and (b) differential Auger spectra for copper.

The energy resolution attainable with the form of CMA shown in Fig. 2.4, although adequate for AES, is not sufficient for XPS. This form of the analyser is also best suited to a point source of electrons rather than the large area source of X-irradiation. This shortcoming can be circumvented, to a large extent, by arranging two analysers in series—the *double pass* CMA. For XPS applications retarding grids are also installed in front of the entrance to the CMA to allow the passage of photoelectrons at constant spectral resolution. When operated in this configuration in the direct energy mode, XPS data of acceptable energy resolution can be obtained. When operated in the usual AES mode, these grids are earthed. For high spatial resolution Auger microscopy, the low flux of Auger electrons means that there is little alternative but to record spectra in the direct energy mode. High spectral resolution Auger spectroscopy is an area of increasing endeavour, and the direct energy spectrum is the obvious choice for this as well. However, the HSA is better suited to these alternatives and there is evidence that it may become the universal analyser for electron spectroscopy over the next few years.

## The hemispherical sector analyser

The schematic diagram of Fig. 2.2 shows a typical HSA configuration for XPS. The hemispherical sector analyser is combined with an electron pre-retardation stage to achieve the resolution required for XPS, and in the schematic of Fig. 2.2 retardation occurs immediately before the electrons enter the hemispheres at the slit plate. This combination may be operated in either the fixed analyser transmission mode (FAT), also known as the constant analyser energy mode (CAE), or constant retard ratio (CRR), also known as fixed retard ratio (FRR), and constant resolving power. In the CAE mode of operation a constant voltage is applied across the hemispheres allowing electrons of a particular energy to pass between them; thus the resolution of the spectrum is constant across the entire energy range. The analyser pass energy is most commonly set at 10, 20, 50, or 100 eV. This is the usual mode of operation for XPS. Referring to Fig. 2.2, electrons are emitted from the specimen and transferred to the focal point of the analyser by the lens assembly. At this point they are retarded electrostatically before entering the analyser itself, (although retardation may occur, according to instrument design, at the mid-point of the lens or be achieved by the lens itself). Those electrons whose energy now matches the pass energy of the analyser are transmitted, detected, and counted by the electron detector and its associated electronics. A complete electron energy spectrum is recorded, either in analogue or digital form, by ramping the retarding field potential and plotting a spectrum of electron energy versus electron counts.

An alternative mode of operation, called constant retard ratio is not so popular for XPS but is widely employed in AES. In the CRR mode of operation, the voltages on the hemispheres of the analyser are ramped with the energy of the spectrum so that the ratio of electron kinetic energy to pass energy is a constant; this constant is the retard ratio of the analyser. The most commonly employed values are CRR = 2, 4, 10, and 20.

Thus with a retard ratio of 10 the pass energy at 100 eV will be 10 eV while at 1000 eV it increases to 100 eV. This mode of operation is particularly popular for AES as the analyser transmission is low at small values of the pass energy, thus effectively suppressing the high electron yield at the low energy end of the spectrum. As both spectral resolution and transmission change with electron energy, quantification of XPS spectra becomes difficult in the CRR mode of operation and CAE is generally preferred. It does, however, have the advantage of accentuating XPS peaks at the low binding energy end of the spectrum. Figure 2.6(a) shows the XPS spectrum of copper

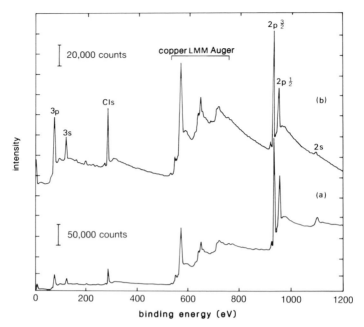

Fig. 2.6. XPS spectrum of copper recorded in (a) the constant resolution mode (CAE = 50 eV) and (b) the constant relative resolution mode (CRR = 20). The analyser transmission is the same for both spectra at a binding energy of 500 eV.

recorded in CAE mode at 50 eV pass energy, whereas Fig. 2.6(b) illustrates the CRR mode at a retard ratio of 20. This gives the same analyser transmission at the mid-point of the spectrum. The change in the ratio of major (Cu $2p_{3/2}$) to minor (Cu 3p) peaks can be clearly seen.

The acquisition of differential spectra is easily achieved with an HSA by applying the a.c. modulation to both hemispheres. Phase sensitive detection is then achieved using a lock-in-amplifier and the channeltron in the current amplification (rather than pulse counting) mode as with the CMA.

In recent years, high resolution Auger microscopes have become popular, and even with high-brightness electron sources, the incident electron flux may not be sufficient to allow phase-sensitive detection. Thus there has been a move towards pulse-counted Auger spectra, the HSA has become increasingly popular for AES, and direct spectra are being published more frequently in the literature. It is of course possible to treat a direct Auger spectrum with an appropriate numerical algorithm to provide a differential output, and most datasystems supplied for electron spectroscopy will include such a routine.

## 2.6 Small area XPS

As already discussed it is possible to manufacture a microfocus X-ray source but this is an expensive route to a small area XPS and also places severe constraints on the amount of space around the specimen available for additional sources and detectors. (A microfocus X-ray monochromator will generally be much smaller than the traditional monochromator to ensure minimum dispersion of the X-rays.) An alternative method is to flood a large area of the specimen with X-rays in the usual manner and to place an aperture between analyser and specimen to restrict the area of analysis. If the spectrometer includes a transfer lens (as shown in Fig. 2.2) of magnification $M$, the area can be reduced still further to $d/M$ (where $d$ is the size of the aperture between lens and analyser). An adjustable aperture may also be incorporated at the bottom end of the transfer lens but this merely defines image resolution; it is not used to define the magnitude of the analysis area.

A logical extension of this method is to use small area XPS in the scanning mode to build up images of the specimen. This may be achieved by the addition of stepper motors to the manipulator or the use of pre-lens electrostatic scanning deflectors so that the specimen remains static but the small region that the analyser 'sees' is deflected. When operated in the latter manner we have an 'XPS microscope', although the resolution is limited by the small area XPS facility; at present this is about 100 $\mu$m, but this may ultimately be improved to perhaps 10 $\mu$m.

At the time of writing, two new approaches to imaging XPS have recently become commercially available; both with demonstrated spatial resolution better than 10 $\mu$m. In the first approach, the specimen is located at the centre of an extremely powerful superconducting magnet: as the photoelectrons are emitted they follow the magnetic lines of force to an imaging device such as a phosphor screen where, as a result of the divergent magnetic field, they reconstruct an enlarged image of the surface chemistry. This type of device originally known as the photoelectron spectro-microscope (PESM), was developed at the University of Oxford for use with ultraviolet radiation, and it seems likely that it will become eminently suitable for use in conjunction with synchrotron beam lines. The other device is similar in concept to the traditional spectrometer of Fig. 2.2 but with the addition of extra lenses between both the transfer lens and analyser, and the analyser and a position sensitive detector. These special lenses convert the spatial information obtained at the specimen into a form which allows the energy of the emitted electrons to be analysed in the usual way as they pass through the analyser. The lens at the output stage simply reconstructs the spatial information at a suitable position sensitive detector, thus providing a two dimensional XPS image of the specimen surface.

## 2.7 Datasystems for electron spectroscopy

There are two primary functions of the datasystems now encountered so frequently on XPS and AES systems; these are control of the spectrometer during acquisition of the spectrum, and the subsequent processing of the data.

The advantages of a datasystem are now well established and such a system will invariably control the spectrometer during acquisition and hold the resultant spectrum in some easily accessible format on floppy or hard disc. One of the main advantages, on the acquisition side, is that weak spectral features can be collected over extended periods of time to allow spectra with improved counting statistics to be obtained. Once recorded on disc, the data may be readily recalled for examination or further processing.

The power of a datasystem does not stop here. The advent of multi-channel electron detectors, especially popular in instances where the electron yield is generally low (such as high spatial resolution AES, small area XPS, and monochromated XPS), has brought about an increased popularity of datasystems and they now become an essential requirement. Ion, X-ray, and electron sources may also be controlled by computer, leading to automatic sputter depth profiling and scanning Auger microscopy, which gives improved experimental data compared with that obtained using analogue control. Computer control enables Auger maps to be recorded digitally and then processed to remove topography effects, usually using a (peak-background)/(background) algorithm. This is far superior to the analogue alternative of using an oscilloscope and a manually established discriminator level.

A more recent innovation is the use of stepper motors to control specimen position, which may enable an analysis to be carried out at different points on the same sample or on different samples using a carousel or *x*–*y* table. Angular resolved XPS may also be carried out under computer control.

Once the spectra have been obtained, some form of dataprocessing is generally required. In AES, this may be limited to quantification of the spectrum, and perhaps differentiation if it was recorded in the pulse counted mode.

In XPS, some form of peak area calculation will be required, together with smoothing and differentiation of the original data, removal of X-ray satellites (Mg and Al radiations both have a series of minor satellites, i.e. the $K\alpha_{3,4}$, $K\alpha_{5,6}$, and $K\beta$ components), and deconvolution and peak fitting routines. An example of a complex spectrum from an argon ion etched molybdenum bearing stainless steel is shown in Fig. 2.7, without the curve fitting routine it would have been extremely difficult to assign the $Mo^{6+}$ components accurately as they are convoluted with the Ar 2p X-ray satellites. Often such routines become exceedingly complex and must be run on a mainframe computer rather than the spectrometer's dedicated datasystem. For these tasks the datasystem must be interfaced to the mainframe, a relatively straightforward task aided by the adoption of a standard data-format by most instrument manufacturers. Such a 'mainframe datasystem' can run routines which enable, for example, complex peak fitting, reconstruction of

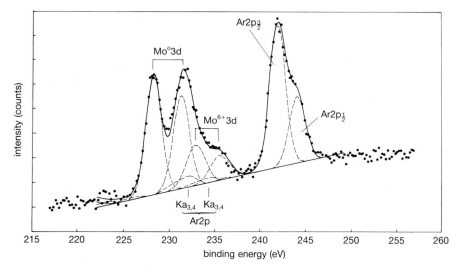

Fig. 2.7. Curve fitting of an ion-etched molybdenum containing stainless steel. The Mo 3d envelope contains 6 individual components (4 from the Mo 3d level and 2 from the AlK$\alpha_{3,4}$ excited Ar 2p level) which must all be accounted for accurately in the peak fitting procedure.

an angular depth profile from a postulated depth profile, correction of sputter depth profiles for instrumental and matrix effects, and manipulation of Auger and X-ray maps.

The other great advantage of a main-frame system is that it allows access to all the peripheral features such as high resolution plotters, high speed printers and archive facilities. However, with the increasing power of mini-computers and workstations it seems likely that 'in-house' datasystems may soon be built around this type of hardware, handling spectra and images from perhaps several spectrometers each equipped with a dedicated data-system supplied by the manufacturer.

# 3 The electron spectrum: qualitative and quantitative interpretation

The product of the electron spectrometer is amenable to many levels of interpretation, ranging from a simple qualitative assessment of the elements present to a full-blown quantitative analysis complete with assignments of chemical states, and determination of the phase distribution for each element. In practice, a happy medium is usually required with an estimation made of the relative amounts of each element present. As there are certain similarities in the way that AES and XPS spectra are treated we shall consider them together as this also provides a means of comparing the analytical capabilities of the two methods.

## 3.1 Qualitative analysis

The first step to be taken in characterizing the surface chemistry of the specimen under investigation is the identification of the elements present. To achieve this it is usual to record a survey or wide scan spectrum over a region that will provide fairly strong peaks for all elements in the periodic table. In the case of AES, a range of 0–1100 eV on a kinetic energy scale is used whereas in XPS 0–1000 eV on a binding energy scale is sufficient. The individual peaks may be identified with the aid of data in tabular or graphical form as reproduced in Appendices 1 and 2, or the *Handbooks of Auger and X-ray photoelectron spectroscopies* published by The Perkin Elmer Corporation. A typical differential Auger spectrum is shown in Fig. 3.1(a). The peaks caused by the elements present, in this case Al, O, C, are observed superimposed on a high background. Auger spectra may be recorded in either the direct (Fig. 3.1(b)), or the differential mode, and nowadays the direct mode appears to be becoming rather more popular with the advent of high spatial resolution scanning Auger microscopes. The photoelectron spectrum from a similar specimen, Fig. 3.1(c), is composed of the individual photoelectron peaks and the associated Auger lines resulting from the de-excitation process following photoemission. Unlike the Auger spectrum of Fig. 3.1(b) the electron background is relatively small and increases in a step-like manner after each spectral feature. This is a result of the scattering of the characteristic Auger or photoelectrons by the matrix bringing about a loss of kinetic energy. The shape of this background itself contains valuable information and to the experienced electron spectroscopist provides a means of

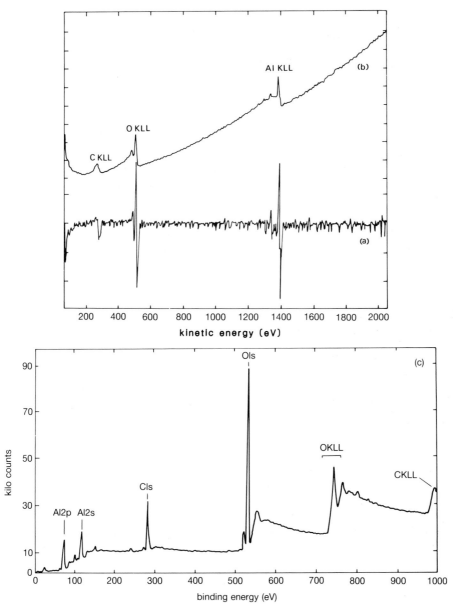

Fig. 3.1. Differential (a) and direct (b) Auger spectra, and the XPS spectrum (c), of oxidized aluminium.

assessing the way in which near-surface layers are arranged. In the case of a perfectly clean surface, the photoelectron peaks will have a horizontal background or one with a slightly negative slope; if the surface is covered with a thin overlayer the peaks from the buried phase will have a positive slope, in the most severe case the peak itself will be absent and the only indication will be a change in background slope at the appropriate energy.

The XPS spectrum is further complicated by the presence of several features of no analytical use such as X-ray satellites and X-ray ghosts. X-ray satellites are present if unmonochromated radiation is used (as is often the case), and occur a constant distance below the main photoelectron peaks; they occur because the characteristic transitions are excited by a minor component of the X-ray spectrum e.g. $AlK\alpha_{3,4}$, $AlK\alpha_{5,6}$, $AlK\beta$. Such features are always present and only present difficulties if they fall at the same binding energy as an element present in very small quantities. The solution may then be to change to another radiation ($MgK\alpha$) as the separations are slightly different. The $AlK\alpha_{3,4}$ X-ray satellite is easily identified in the XPS spectrum of Fig. 3.1(c) for the most intense photoelectron peak (O 1s), as a small peak at a binding energy of approximately 520 eV. Auger transitions present in an XPS spectrum do not show such satellite features and this provides a rapid means of distinguishing between the two as seen on the $OKL_{2,3}L_{2,3}$ peak. X-ray ghosts arise from unsuspected X-rays irradiating the sample, these may result from 'crosstalk' in a twin anode gun, (the generation of a small amount of characteristic X-radiation as a result of anode misalignment, from the anode material not in use in addition to the X-ray flux from the chosen source, or possibly $CuL\alpha$ ($h\nu = 929.7$ eV) from the exposed base of a damaged anode. In either case the problem should be reduced to an inconsequential level by overhauling and readjusting the X-ray gun.

## 3.2 Chemical state information

The ability to detect an element, not only in very thin layers, but also to assign its valence state is extremely useful and it is at this point that the two techniques start to diverge slightly. One of the major strengths of XPS is its ability to provide this information and various data sets exist that catalogue the XPS chemical shift. A common occurrence is the convolution of two or more chemical states in a single spectrum, as illustrated in Fig. 2.7. This situation is frequently encountered in the surface analysis of polymers and the examination of very thin passive films on metals as we shall see in the following chapters.

In Auger electron spectroscopy, the situation is not so promising. This is because the Auger transition involves three electrons, two of which are often in the band structure of the material. This leads to very broad poorly defined peaks for most elements, and assignment of chemical state tends to rely on the shape of the peak rather than position, as shown in the $OKLL$ spectra for a series of organic compounds in Fig. 3.2(a). If the two outer electrons are not valence electrons a sharp peak may result as we observe for the $KLL$ series of peaks of aluminium and silicon, and the $LMM$ series of copper, zinc, gallium, germanium, and arsenic. The X-ray excited Ge$LMM$ Auger spectra of Fig. 3.2(b) show components attributable to $Ge^0$ and $Ge^{4+}$

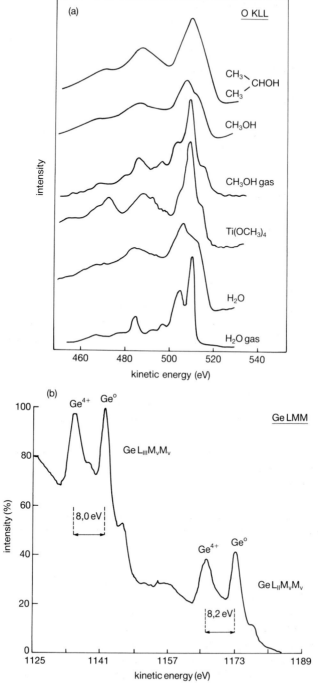

Fig. 3.2. Auger chemical state information for (a) a series of organic compounds and water, and (b) a germanium single crystal with a thin layer of oxide.

separated by over 8 eV. The reason Auger spectroscopy is generally credited with an inability to detect chemical information springs from the poor spectral resolution of the early Auger electron analysers; the information was often there but the analysers were unable to detect it! It is, however, very difficult to employ Auger spectroscopy to resolve mixtures of chemical states as the peak fitting requirements become extremely complex.

The cornerstone of any spectral analysis which relies on peak position to provide information presupposes the ability to determine such values with the necessary accuracy, at least $\pm 0.1$ eV. The two possible sources of error are those due to spectrometer calibration and those resulting from electrostatic charging of the specimen. The former are easily overcome by accurate calibration of the spectrometer against known (standard) values for copper and gold. The latter are resolved for metallic specimens by proper mounting procedures but for insulators and semi-conductors a slight shift as a result of charging is always a possibility. Although it is possible to use an internal standard such as the adventitious carbon 1s position, this is not particularly accurate and will vary slightly with the form and amount of carbon. A much more attractive method is to make use of the chemical shift on both the Auger and photoelectron peak in a XPS spectrum and to record the separation of the two lines, this quantity is known as the Auger parameter ($\alpha$) and is numerically defined as the sum of the peaks minus the photon energy;

$$\alpha = E_B + E_K - h\nu$$

where $E_B$ is the binding energy of the photopeak and $E_K$ is the kinetic energy of the Auger transition. The measured value will thus be independent of any electrostatic charging of the specimen. The elements that yield useful Auger parameters in conventional (AlK$\alpha$) XPS include F, Na, Cu, Zn, As, Ag, Cd, In, and Te, when using high energy XPS the list can be extended to include Al, Si, P, S, and Cl. As well as providing chemical state information the Auger parameter is, in some cases, able to provide information on crystal structure and relaxation energies.

Within the high resolution spectra of individual core levels there may exist fine structure that gives the electron spectroscopist additional information concerning the chemical environment of an atom. The major features in this category are multiplet splitting and 'shake-up' satellites.

### 3.2.1 Shake-up satellites

Shake-up satellites occur when the outgoing photoelectron simultaneously interacts with a valence electron and excites it (shakes it up) to a higher energy level; the energy of the core electron is then reduced slightly giving a satellite structure a few electron volts below (but above on a binding energy scale) the core level position. Such features are fairly rare, the most notable examples being the 2p spectra of the d-band metals and the bonding to anti-bonding transition of $\pi$ molecular orbital electrons ($\pi \rightarrow \pi^*$ transition) in aromatic organics. The former is best illustrated by the Cu 2p spectrum; a strong shake-up satellite is observed for CuO, as shown in Fig. 3.3(a), but is absent for $Cu_2O$ and metallic copper. An allied feature is the 'shake-off' satellite where the valence electron is ejected from the ion completely. These are rarely seen as discrete features of the spectrum but more usually as a broadening of the core level peak or contributions to the inelastic background.

### 3.2.2 Multiplet splitting

Multiplet splitting of a photoelectron peak may occur in a compound that has unpaired electrons in the valence band, and arises from different spin distributions in the electrons of the band structure. This results in a doublet of the core level peak being considered; multiplet splitting effects are observed for Mn, Cr, (3s levels) Co, Ni ($2p_{3/2}$ levels), and the 4s levels of the rare earths. The $2p_{3/2}$ spectrum of nickel shows multiplet splitting for NiO, as shown in Fig. 3.3(b), but not for $Ni(OH)_2$—a feature that has proved very useful in the examination of passive films on nickel. The final type of loss feature to be considered is that of plasmon losses. These occur in both Auger and XPS spectra and are specific to clean metal surfaces. They arise when the outgoing electron excites collective oscillations in the conduction band electrons and thus suffers a discrete energy loss (or several losses in multiples of the characteristic plasmon frequency, about 15 eV for aluminium). One of the characteristic plasmon loss peaks for clean aluminium is shown in Fig. 3.3(c).

The loss features described above can provide valuable information but some, as in the case of plasmons, merely serve to complicate the spectrum. In either case, it is important to assign them correctly so that all spectral features are accounted for and all elements identified before beginning the calculation for a quantitative surface analysis.

## 3.3 Quantitative analysis

There are two basic approaches that may be taken in carrying out the calculations for a quantitative evaluation of surface composition:

(1)  based on first principles;

(2)  based on an empirical relationship together with cross-sections or sensitivity factors which may be published or determined in house.

As the quantification of an XPS spectrum is rather more straightforward and potentially more accurate we shall consider it first.

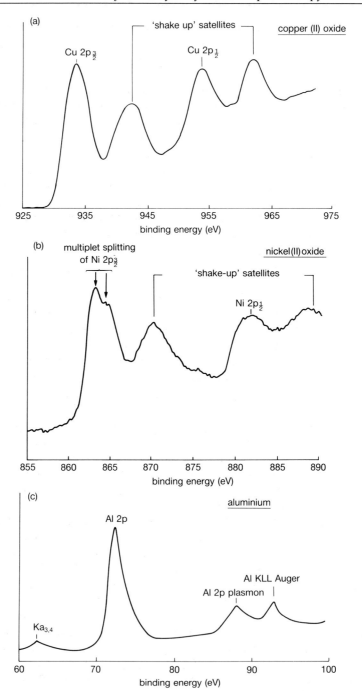

Fig. 3.3. Satellite features present in XPS. (a) Shake-up satellites in CuO, (b) multiplet splitting in NiO, (c) plasmon losses from clean aluminium.

The intensity ($I$) of a photoelectron peak from a homogeneous solid is given, in a very simplified form, by:

$$I = J\rho\sigma K\lambda$$

where $J$ is the photon flux, $\rho$ is the concentration of the atom or ion in the solid, $\sigma$ is the cross-section for photoelectron production (which depends on the element and energy being considered), $K$ is a term which covers instrumental factors such as analyser transmission function and detector efficiency and $\lambda$ is the electron inelastic mean free path (see Section 1.4). The intensity referred to will usually be taken as the integrated area under the peak following the subtraction of a linear or S-shaped background. The above equation can be used for direct quantification (the so-called first principles approach) but more usually experimentally determined sensitivity factors, ($F$), will be employed. The parameter $F$ includes the terms $\sigma$, $K$, and $\lambda$, in the standard equation, as well as additional features of the photo-electron spectrum such as characteristic loss features. Once a set of peak areas has been calculated for the elements detected, $I$ in the above equation has been determined. The terms $\sigma$, $K$, and $\lambda$ are incorporated into a set of sensitivity factors appropriate for the spectrometer used. If the X-ray flux remains constant during the experiment, (as it invariably does), we can determine the atomic percentage of the elements concerned, by dividing the peak area by the sensitivity factor and expressing as a fraction of the summation of all normalized intensities:

$$[A] \text{ atomic } \% = \{(I_A/F_A)/\Sigma(I/F)\} \times 100\%.$$

The calculation of surface composition by this method assumes that the specimen is homogeneous within the volume sampled by XPS. This is rarely the case but even so, the above method provides a valuable means of com-paring similar specimens. For a more rigorous analysis angular dependent XPS may be employed to ensure lateral homogeneity and to elucidate the hierarchy of overlayers present.

The quantitative interpretation of Auger spectra is not as straightforward. The first problem encountered is the form of the spectrum. In the differential mode the intensity measurement is the peak-to-peak height. For low resol-ution spectrometers this is approximately proportional to peak area; for high resolution studies, fine structure which becomes apparent in the spectrum reduces the apparent peak-to-peak height. It is for this reason that the integrated peak area of the direct energy spectrum is often preferred for quantitative AES. The relative peak areas in a spectrum will depend on the primary beam energy used for the analysis and also the composition of the specimen. It is the latter, matrix, effect that has prevented the production of a series of AES sensitivity factors of the type widely used for XPS. Instead, it is necessary to fabricate binary or ternary alloys and compounds of the type under investigation to provide calibration by means of a similar Auger spectrum; however, the sensitivity factors produced have a narrow range of applicability. In this manner, it is possible to determine the concentration of an element of interest ($N_A$) as follows:

$$N_A = I_A/(I_A + F_{AB}I_B + F_{AC}I_C + \ldots)$$

where $I$ is the measured intensity of the element represented by the subscript, and $F$ is the sensitivity factor determined from the binary standard such that:

$$F_{AB} = (I_A/N_A)/(I_B/N_B)$$

Various semi-quantitative methods are employed by laboratories throughout the world which relate a measured Auger electron intensity to that of a standard material under the same experimental conditions and this seems to be a fairly satisfactory approach where the time and expense of producing the relevant standard specimens is not warranted.

Although a surface analytical study may be an end in itself, knowledge of the concentration of elements near to the surface is often required. To achieve this, some form of compositional depth profiling is required, either by destructive or non-destructive means and this adds another degree of complexity to the interpretation of the resultant spectra as we shall see in the next chapter.

# 4 Compositional depth profiling

Although both XPS and AES are essentially methods of *surface* analysis it is possible to use them to provide compositional information as a function of depth. This can be achieved in two ways; by manipulating the Beer–Lambert equation of Section 1.4 either to increase or decrease the integral depth of analysis (by changing the geometry of the experiment, or the energy of the emitted electron and hence the inelastic mean free path). Alternatively, material from the surface of the specimen can be removed either *in situ* (by ion sputtering) or before the specimen is introduced into the spectrometer. Analysis is then carried out sequentially with material removal and a compositional depth profile gradually built up.

## 4.1 Non-destructive depth profiling methods

These methods are used almost exclusively in photoelectron spectroscopy. Although the principles are equally applicable to Auger electron analysis, the results obtained with the high lateral resolution employed in AES and SAM mean that such changes in analysis depth occur in the analysis of parts of the specimen with different orientations to the electron analyser (because of specimen surface roughness). Such effects tend to be regarded as experimental artefacts to be circumvented by the Auger microscopist, and have only recently become the subject of rigorous scientific investigation.

If we consider the Beer–Lambert equation discussed in Chapter 1, it is clear that the depth of analysis is dependent on the electron take-off angle $\theta$. By turning the specimen to a low value of $\theta$, say 15°, an analysis is recorded which is extremely surface sensitive. As normal electron emission is approached ($\theta = 90°$) so the analysis depth moves towards the limiting value of $3\lambda$. This value is often referred to as the XPS analysis depth although it is, of course, more correctly described by $3\lambda \sin \theta$. The relative sampling depths at different take-off angles are illustrated schematically in Fig. 4(a). A thin overlayer will give a characteristic angular distribution predicted by the Beer–Lambert expression, as shown in Fig. 4.1(b). An island-like distribution will show no angular dependence, thus it is possible to distinguish between these two types of phase distribution with relative ease.

This manner of depth profiling is invaluable for compositional changes that occur very close to the surface and has been employed for studies of thin passive films on metals and surface segregation in polymers. Polymers present some special problems in surface analysis in that they are not

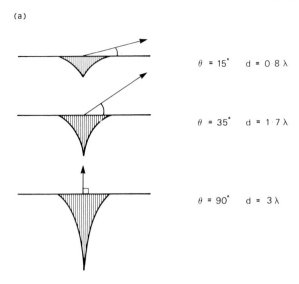

where λ is the electron inelastic mean free path

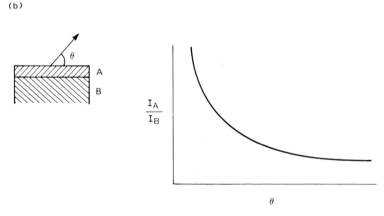

Fig. 4.1. Angular electron emission (a) sampling depth as a function of electron take-off angle (θ), (b) substrate/overlayer and intensity versus θ.

generally amenable to analysis by AES or ion beam compositional depth profiling because of sample charging and degradation problems. Consequently angular resolved XPS is one of the few ways of probing near-surface compositional gradients. A reconstructed angular depth profile from the failure surface of a thin organic coating detached from its steel substrate, obtained by this method is shown in Fig. 4.2.

An alternative way of obtaining in-depth information in a non-destructive manner is by examining electrons from different energy levels of the same

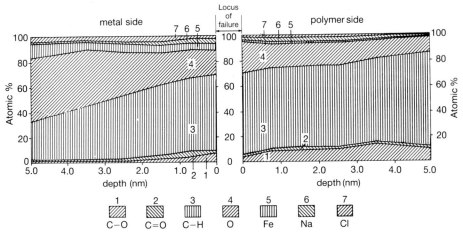

Fig. 4.2. Reconstructed depth profile from angle resolved XPS data. The diagrams show the changes in composition away from the locus of failure of a polymer coating peeled from a steel substrate. In the depth profile the shaded areas represent the changes in the relative abundance of each chemical type as a function of depth.

atom. The inelastic mean free path varies with kinetic energy, and by selecting a pair of electron transitions that are both accessible in XPS but have widely separated energies, it is possible to obtain a degree of depth selectivity. The Ge 3d (kinetic energy = 1450 eV, electron IMFP = 1.6 nm) spectrum of Fig. 4.3(a) shows $Ge^0$ and $Ge^{4+}$ components with the oxide component being about 80 per cent of the elemental. If we also record the spectrum of the Ge $2p_{3/2}$ region (kinetic energy = 260 eV, $\lambda = 0.7$ nm) we see the elemental component is merely a shoulder on the $Ge^{4+}$ peak, Fig. 4.3(b), thus confirms the presence of the oxide layer as a surface phase. It is possible to obtain a similar effect by using the same electron energy level but exciting the photoelectrons with a series of X-rays. For instance, C1s electrons have a constant binding energy of 285 eV, which corresponds to a kinetic energy of 969 eV in MgK$\alpha$ radiation, 1202 eV in AlK$\alpha$, and 2700 eV in AgL$\alpha$ radiation. These kinetic energies yield analysis depths of approximately 6, 7, and 10 nm for a typical polymer. Although high energy X-ray sources for XPS are still rare, the conventional Al/Mg twin anode fitted to most spectrometers does provide a modest depth profiling capability which is often sufficient to distinguish between a surface layer and an island distribution.

## 4.2 Depth profiling by erosion with inert gas ions

Although the non-destructive methods described above are extremely useful for assessing compositional changes in the outer 5–10 nm of a material, to go further than this, it is necessary to remove material by ion bombardment

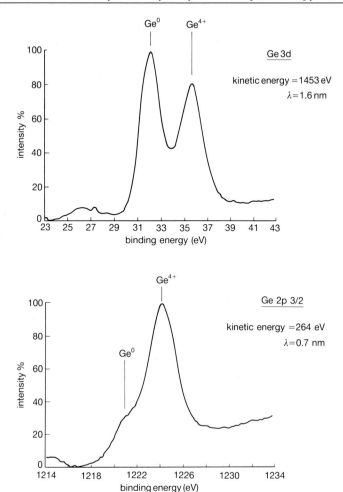

Fig. 4.3. Ge3d and Ge2p$_{3/2}$ spectra showing variation of sampling depth with electron kinetic energy.

within the spectrometer. Unless the analysis being carried is very rapid (perhaps just 2 or 3 Auger peaks being recorded) the ion bombardment is carried out in a sequential manner with the ion gun switched off, usually automatically by computer, while the spectrum is recorded. The resulting data is presented as elemental intensity versus etch time, and a major problem in sputter depth profiling is converting this etch time scale to a depth scale. Although, in special cases, it is possible to calibrate ion guns for a particular material, it is a time consuming procedure and more often the sputter rate is related to an international standard. The current standard is a $Ta_2O_5/Ta$ foil with an accurately determined oxide thickness (30 nm or 100 nm). Thus it is possible to report a sputter rate and also an interface width for the ion gun and the conditions used in any particular piece of work. In addition to providing a means of compositional depth profiling in surface analysis, ion guns may be fitted to a spectrometer for additional purposes such as large area specimen cleaning or as the primary beam in ion beam analysis methods such as ISS or SIMS. The requirements for each application are slightly different and there are several different designs in common use.

### 4.2.1 Ion gun design

The three most widely employed types of ion source for surface science are, in order of increasing performance and cost, the cold cathode static spot gun, the electron impact ion source, and the duoplasmatron type of ion gun.

In the cold cathode type of gun a variable potential of between 1–10 kV is utilized in conjunction with an external magnet to produce a discharge in the ionization region of the gun which has been back-filled with argon or another inert gas to a pressure of about $10^{-6}$ mbar. The positively charged ions are extracted and the beam shaped by a simple focus electrode and these guns generally give a static spot size of 5–10 mm depending on the applied focus potential. They are recommended for specimen cleaning, but with the addition of an aperture assembly below the focus electrode, they can give remarkably uniform etch craters. The main disadvantage is the production of neutral species which are not deflected by the focus potential and give rise to a 'sub-crater' of about 5–10 per cent of the area of the main crater. The maximum current available with this type of gun is of the order of 50 $\mu$A. Cold cathode ion guns of this type are used for specimen cleaning, and, because of their large spot size, XPS depth profiling.

For the rapid removal of material and etching of large areas the duoplasmatron design of ion source is sometimes preferred. A magnetically constricted arc is used to produce a dense plasma from which the ion beam is extracted, focused, and rastered across the specimen by a set of deflector plates. Duoplasmatron ion sources invariably use cold cathode discharge, and although hot cathode electron impact designs are theoretically possible, the cold cathode is preferred, as it allows reactive gases such as hydrogen, oxygen, and nitrogen to be used for special applications. Ion guns based on a duoplasmatron source provide an intense source with a narrow energy spread, making it suitable for small spot focusing. The current density achieved with such an ion gun can be high leading to high etch rates. Duoplasmatron ion sources are available in a range of spot sizes varying from 2 mm (providing 80 $\mu$A of ion current and with a field of view of approximately 15 mm × 15 mm) to better than 5 $\mu$m (providing a maximum current of 5 $\mu$A and a 2 mm × 2 mm field of view). The former is ideal for large area depth profiling in XPS whilst the latter finds extensive use as a primary source in the ion beam analysis of materials.

The alternative to a cold cathode for the production of ions is the electron impact source. In this type of source, electrons from a heated filament are accelerated into a cylindrical grid where they collide with gas atoms giving rise to the formation of ions. The ion energy is controlled by the magnitude of the potential applied to the grid (up to 5 kV). This ion source produces a narrow energy spread with a small fraction of neutrals (i.e. unionized atoms) present in the beam. Spot sizes commercially available vary from 2 mm (5 $\mu$A) down to 50 $\mu$m (500 nA). This type of ion source is very popular for AES depth profiling and the beam will often be rastered over the sample surface to produce a square etch pit, although it can, of course, be used in the static mode. The large area craters required for XPS depth profiling mean the current densities employed (and therefore the etch rates achieved) are low and for this reason cold cathode sources will be used in preference to electron impact sources.

With both duoplasmatron and electron impact sources the beams may be 'purified' (i.e. to remove impurity and multiply charged ions) by the addition of a Wien filter, (a crossed magnetic and electrostatic field mass separator), and a small deflection within the gun design to eliminate neutrals. Such high purity beams are not, however, a prerequisite for good quality XPS and AES sputter depth profiling. In general, the only precautions necessary are the provision of a high purity gas feed, together with a titanium getter near the gas admission valve, to ensure the removal of residual oxygen which would otherwise present problems in the depth profiling of reactive metals such as chromium or titanium.

### 4.2.2 Sputter depth profiling

The use of ion beams to remove material during surface analysis in order to produce a compositional depth profile is an attractive proposition but there are certain artefacts which may lead to difficulties or errors in interpretation of the results. These can be divided into

(1)  features arising from the interaction of the ion beam with the solid;

(2)  the manner in which the experiment is carried out; and

(3)  features present on the specimen under investigation.

Dealing with ion beam-solid interactions first, the literature available on this subject is enormous (see bibliography for examples) and all that can be achieved here is to make the reader aware of possible causes of profile distortion. The primary process is that of sputtering surface atoms to expose underlying atomic layers. At the same time, some of the primary ions are implanted into the substrate and will appear in subsequent spectra. Atomic (cascade) mixing results from the interaction of the primary ion beam with the specimen and leads to a degradation of depth resolution. Enhanced diffusion and segregation may also occur and will have the same effect. The sputtering process itself is not straightforward, there may be preferential sputtering of a particular type of ion or atom. Ion-induced reactions may occur; for instance copper (II) is reduced to copper (I) after 10–15 seconds in a low energy ion beam, i.e. at very low ion dose. As more and more material is removed so the base of the etch crater increases in roughness and eventually interface definition will become very poor indeed.

Contributions arising from the ion beam may be significant but instrumental factors also play a large part in the quality of the sputter profile. If the quality of the vacuum is poor or the specimen very reactive (e.g. metallic titanium) the surface which is analysed may not be the same as the material composition owing to reaction with residual gases of the analysis chamber; similarly, care must be taken to avoid the redeposition of sputtered species onto samples awaiting analysis. The etch crater must be coincident with the analysis area and, particularly in XPS, the crater walls must be avoided. The type and design of ion gun may play an important role and, in general, impurities and neutrals are to be avoided; in particular the gas must be free from oxygen which may be achieved by passing over hot titanium sponge. Controlling the intensity and stability of the ion beam is also essential and these parameters are best optimized for the system concerned. This is achieved by undertaking a series of depth profiles on a standard material supplied for the purpose such as the NPL $Ta_2O_5$/Ta reference material or the NBS Ni/Cr multilayer structure. As the interface width attainable under ideal conditions is documented by the suppliers of the reference materials, the analyst can, by varying experimental parameters, arrive at the optimum conditions for depth profiling with the spectrometer.

The composition and form of the specimen are also important, in particular surface roughness should be kept to a low value. The effect of a very rough surface is twofold. First, depending on the system geometry there may be a problem of shadowing between the ion beam and primary radiation. Secondly, since a rough surface will present facets of differing geometry to the ion beam, so the sputter yield will vary locally within the region being ion etched, as the angle of incidence of the ion beam varies locally leading to uneven etching of the specimen.

In the sputtering of the surface of alloys or compounds, the phenomenon of preferential sputtering may become a problem. This occurs because the sputter yield of one of the component elements is greater than the others, effectively reducing the concentration of that element at the surface until steady state conditions are approached. In the subsequent XPS or AES experiments the quantitative analyses will indicate a depletion, with depth, of that particular element. This is indeed the case, but such depletion results from the sputtering process, and is not an intrinsic property of the specimen itself. Once such phenomena have been identified, corrections in the subsequent compositional depth profile must be made.

However, in spite of the potential problems outlined above, sputter depth profiling is by far the most popular means of producing a compositional depth profile in surface analysis, and it is fair to say that the majority of the problems can be circumvented or reduced to an acceptable level by careful experimental technique. It is for this reason that as a method for depth profiling it is widely used in studies of metals, oxides, ceramics, and semi-conductors, as we shall see in the next chapter.

The analysis depth which is feasible varies with the sample and the system employed but 1–2 $\mu$m is regarded as the upper limit; further than this, it is necessary to resort to an *ex situ*, mechanical process for removing material. Material is removed by polishing the specimen at a very shallow angle (1–3°) and then introducing the specimen, with any buried interfaces now exposed, into the spectrometer. A brief ion etch to remove contamination is all that is needed prior to analysis. By carrying out Auger point analyses in a stepwise manner the variation of concentration with depth is established and it is a matter of simple geometry to convert position of the analysis in the $x$–$y$ plane to distance from the original surface, the $z$ plane. The main difficulty with this technique is the need to produce a very shallow taper section with a flat surface and well defined geometry. To overcome this problem, a method of producing a shallow hemispherical crater has been developed, the so-called ball cratering approach.

## 4.3  Depth profiling by AES and ball cratering

In the ball cratering process, mechanical sectioning of the specimen is carried out by a rotating steel ball (30 mm in diameter) coated with fine (1 $\mu$m) diamond paste which rotates against the specimen and fashions a shallow saucer-like crater. The ball can be removed from time-to-time to assess the progress of the lapping and, on replacement, automatically 'self-centres' in the crater. From a knowledge of the diameter of the sphere ($2R$) and the crater ($D$) the depth ($d$) can be calculated as

$$d = \frac{D^2}{4(2R + d)}$$

but as $d$ is very small compared with $R$ this is approximated to $d = D^2/8R$. By recording Auger analyses along the surface of the crater a compositional depth profile is achieved. If there are buried interfaces of special interest ion sputtering may be used at a point on the crater close to the interface to obtain better depth resolution. Although ball cratering works well for metals and oxides, there are problems with both soft and certain brittle materials. Polymers are extremely difficult to handle but some success has been obtained recently by using a ball cratering machine equipped with a cryo-stage.

## 4.4 Conclusions

The provision of an extended surface analysis or compositional depth profile is something that will be asked of most practising surface analysts at some-time or other. Providing this information may be straightforward, for example the examination of a 4 nm passive film on stainless steel; or extremely taxing, for example the investigation of a polymer metal interface buried between many micrometres of polymer and oxide. The methods outlined in this chapter can be combined with complex specimen prep-aration methods to provide the required information, and it is inevitable that the successes of the last 15–20 years are to be surpassed by the development of new combinations of techniques for depth profiling. These will probably concentrate on rapid methods of material erosion to give taper sections across interfaces that can subsequently be analysed fairly quickly, thus making the most efficient use of instrument time. Initial results from ion beam bevelling of specimens and *in situ* machining of taper sections both look very promising.

# 5 Applications of electron spectroscopy in materials science

## 5.1 Introduction

So far, in this text, we have been concerned with the practice of electron spectroscopy and the interpretation of the resultant spectrum. We will now consider the way in which it is possible to make use of these surface analysis techniques to provide information which furthers our knowledge in a particular discipline. Although XPS and AES together with SAM are used widely in all branches of pure and applied sciences—as well as for trouble-shooting and quality assurance purposes—the only area that we will consider in this chapter, is their use in materials science investigations. If we subdivide this group, it is possible to identify the following applications headings; metallurgy, corrosion, ceramics (including mineralogy and catalysis), electronic devices, polymers, and adhesion. We shall consider each of these areas in turn, representative references for each are listed in the bibliography which will provide interested readers with further examples and guidance in their particular field.

## 5.2 Metallurgy

In the field of metallurgy, it is Auger electron spectroscopy that has proved to be the most popular technique, and with good reason. The majority of investigations are concerned with the diffusion of elements within metallic matrices. This may take the form of interdiffusion of metallic coatings with the substrate or of the surface segregation of minor alloying elements on heating in oxidizing or reducing atmospheres. However, the major contribution of Auger electron spectroscopy to metallurgy, especially in the early days of the development of surface analysis, was the investigation of grain boundary segregation and embrittlement in structural steels. In addition, both AES and XPS have been used in 'quality assurance' and sometimes 'forensic' roles to ensure (for example) rolled steel sheet of adequate cleanliness or to identify surface phases which lead to poor compaction in powder metallurgy processing.

The embrittlement of structural steels results from the aggregation of certain elements, present in very low or trace quantities in the bulk material, at the prior austenite grain boundaries. The grain boundaries are weakened to such an extent that they become the preferred fracture path, with catastrophic effects on the material's mechanical properties. The elements most widely investigated are phosphorus and sulphur but the effect is brought about by, and has been studied for, silicon, germanium, arsenic, selenium, tin, antimony, tellurium, and bismuth. As the quantity of grain boundary segregant involved is necessarily very small, probably sub-monolayer, and located at the grain-boundary within a material of grain-size of about 100 $\mu$m or less, the need for surface specificity and reasonable spatial resolution is immediately apparent. In order to measure the quantity of segregant at the interface, the steel must be fractured in an intergranular manner, usually at, or near, liquid nitrogen temperature. This must be carried out within the UHV environment of the spectrometer to prevent oxidation of the iron matrix and subsequent obliteration of the small signal from the segregant. Nowadays, most manufacturers offer such a fracture stage for their Auger microscopes, the more sophisticated having the ability to analyse both fracture surfaces. All rely on fracture by a fast three point bend configuration similar to a Charpy Test. Scientists requiring controlled strain rate fracture must still resort to building their own devices!

The surface morphology generated by such low temperature fracture is sometimes a mixture of regions of both intergranular and transgranular failure. This provides a convenient comparison between the matrix composition (transgranular) and a grain boundary analysis from the region of brittle failure, Auger spectra taken from regions of this type are presented in Fig. 5.1. The presence of oxygen indicates that even under clean UHV

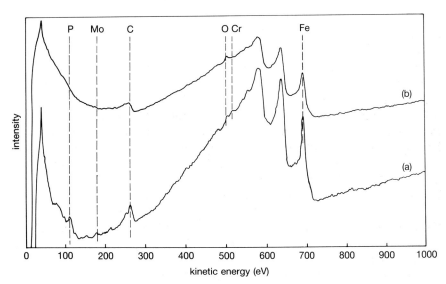

Fig. 5.1. Auger spectra from regions of (a) intergranular and (b) transgranular failure of an alloy steel fractured at liquid nitrogen temperatures.

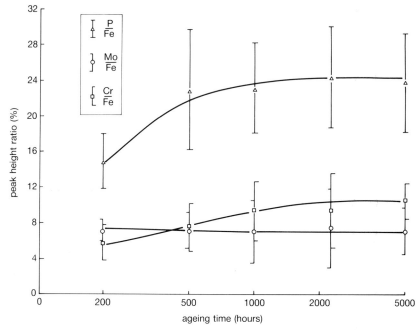

Fig. 5.2. Grain boundary composition versus heat treatment times for an alloy steel.

conditions some contamination of the surface may occur. Further evidence of such contamination is provided by the carbon level obtained, although in the case of the intergranular failure much of the carbon is attributable to grain boundary carbides. It is often informative to compare the degree of grain boundary segregation present in a steel after a series of thermal treatment times; in Fig. 5.2 the elemental ratios (taken as a ratio of raw Auger data) for P/Fe, Mo/Fe, and Cr/Fe, are shown for a 2.25 Cr 1.0 Mo steel as a function of ageing time. The phosphorus segregation reaches a plateau after about 1000 hours as expected, more surprising is the manner in which the Cr/Fe ratio follows the P/Fe trend, this results from carbide formation at grain boundaries. By carrying out compositional depth profiling on the freshly generated fracture surfaces it is possible to confirm that the elements phosphorus and chromium are enriched at the grain boundary. Figure 5.3 shows the results of such a depth profiling experiment from the 1000 hour specimen of Fig. 5.2, the P/Fe and Cr/Fe ratios fall rapidly with ion dose, whereas the Mo/Fe values show relatively little change.

Studies have been made of many systems that exhibit grain boundary segregation and the underlying theory is now well developed, mainly as a result of systematic studies undertaken at the National Physical Laboratory. Thus, the extent of grain boundary segregation may be predicted by the following equation for dilute levels of segregant in the matrix:

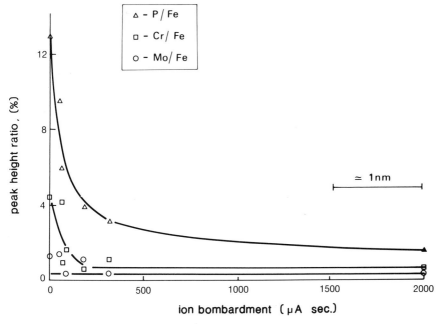

Fig. 5.3. Depth profile at grain boundary for a steel aged for 1000 hours.

$$\beta = \frac{K}{X_c^0}$$

Where $\beta$ is the grain boundary enrichment ratio, $X_c^0$ is the solid solubility of segregant in the matrix and $K = \exp(-\Delta G/RT)$, $\Delta G$ is the free energy of segregation. This equation describes a large number of experiments undertaken on many systems all indicating that the degree of enrichment is dependent on solid solubility over a very wide range (from 100 p.p.m. to 100 per cent). This data is presented graphically in Fig. 5.4.

Thus in the field of segregation and embrittlement AES has not only provided a technique that enables the level of segregant to be qualitatively assessed, it has, as a result of such measurements, enabled the development of an underlying theory which predicts such a phenomenon very accurately.

## 5.3 Corrosion science

There are two areas within corrosion science in which electron spectroscopy has had a dramatic impact; the interaction of a metal surface with its environment, perhaps to form a passivating overlayer, and the breakdown of the surface film by a localized phenomenon such as pitting. The former is readily studied by XPS; where the ability to separate the spectrum of the underlying

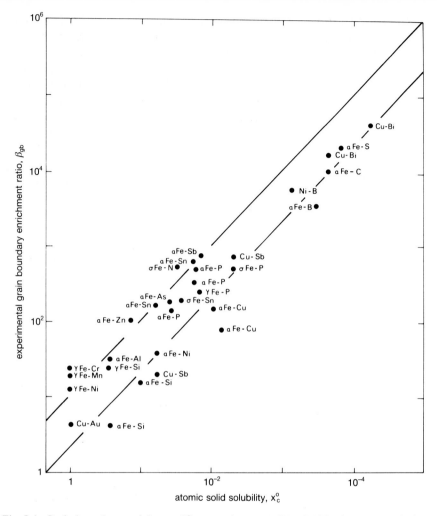

Fig. 5.4. Grain boundary enrichment ($\beta$) versus inverse solid solubility for a range of binary alloys.

metal from that of its oxide film enables the definitive identification of passivating films on alloys. It also enables the thickness of very thin films to be estimated using the Beer–Lambert equation of Chapter 1. However, although XPS provides valuable information concerning the composition and perhaps growth kinetics of such films, their breakdown, leading to major environmental degradation of the metal, is a localized phenomenon requiring high spatial resolution surface analytical methods, i.e. sub-micron scanning Auger microscopy. The provision of compositional depth profiles of corrosion films is a standard requirement and may be achieved by using argon ion bombardment in conjunction with either AES or XPS. In the case of very thin films (<4 nm) angle resolved XPS can be very informative.

Although in some cases the deconvolution of the metallic and cationic spectral information is straightforward, there are cases of great importance to the corrosion scientist where a major research effort is required to unravel the complexities of two or more valence states along with loss features present in the spectrum. The transition metals in general have presented difficulties; in the case of iron the need is to be able to distinguish Fe(II) from Fe(III) confidently, as shown in Fig. 5.5. The two spectra indicate that there is a binding energy difference for approximately 1.5 eV between the $2p_{3/2}$ position of the chemical states; however, it is the position of the shake up

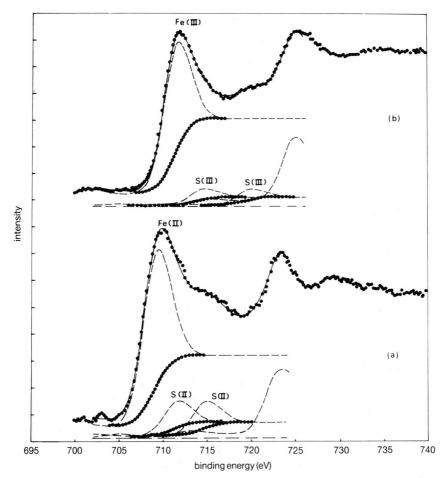

Fig. 5.5. Curve fitting of iron $2p_{3/2}$ spectra of (a) Fe(II) and (b) Fe(III). The satellites (S) provide a valuable aid to assignment of chemical state.

satellites in the valley between the $2p_{3/2}$ and $2p_{1/2}$ that give the most reliable information. In the case of the Fe(II) state, these features give rise to a broadening on the lower binding energy side of the valley (Fig. 5.5(a)), for Fe(III) on the higher energy side, (Fig. 5.5(b)). The curve fitted spectra of Fig. 5.5 show the relative positions and intensities of the satellites. In the case of a mixed phase, all these components, together with Fe(0) if the film is very thin, must be considered, preferably for both 2p peaks as a check for self-consistency. The backgrounds of the individual singlets will also vary with depth distribution. It is only by using very sophisticated computer curve fitting routines, in combination with extensive knowledge of peak shape and relative intensities, that this has become possible. This type of information is well documented for all elements of interest to the corrosion scientist although, in some cases, it is necessary to consider the X-ray induced Auger peaks in the XPS spectrum as well. These are particularly informative for magnesium, copper, and zinc, where the photoelectron spectra alone do not provide unambiguous chemical state information.

A little used, but nevertheless valuable, method in which surface analysis can assist in the determination of electrochemical history, is by the monitoring of cations or anions adsorbed from aqueous solution. For instance, if a metal electrode is polarized cathodically in $MgCl_2$ solution, it will preferentially adsorb cations on the metal surface, seen in the electron spectrum (XPS or AES) as an excess of magnesium to chlorine. If the electrode has been polarized anodically the reverse is true. This approach is useful in establishing if a pit, or other corrosion site, is active or benign, it can also be used to assess the potential distribution around an active pit. An experiment of the latter type is described in Fig. 5.6. The broken line and right hand axis indicates the anion/cation ratio determined on large electrodes as a function of electrode potential for a constant charge passed, plotted against the predicted potential distribution for small pits. Microanalysis of three pits of about 2 $\mu$m in diameter by Auger spectroscopy allow the electrode potential distribution around a pit to be determined experimentally, and show excellent agreement with that predicted by the broken line (the combination of XPS and theory), indicating an active anodic centre to the pit surrounded by a cathodic halo as illustrated in the schematic of Fig. 5.6.

Such pitting may be related to microstructural features of the alloy, and the addition of X-ray analysis facilities to a scanning Auger microscope enables features such as inclusions to be identified and imaged at the same time as the surface phases. In Fig. 5.7 a combination of Auger and X-ray maps indicate that a pit, identified in the SEM image is shown to be active (by the concentration of chlorine in the surface analysis), and associated with a CuS inclusion group, identified by EDX. Figure 5.7 also summarizes the other information of value in corrosion studies, the chemical state information from XPS, and the compositional change within the very thin passive film obtained by sputter depth profiling.

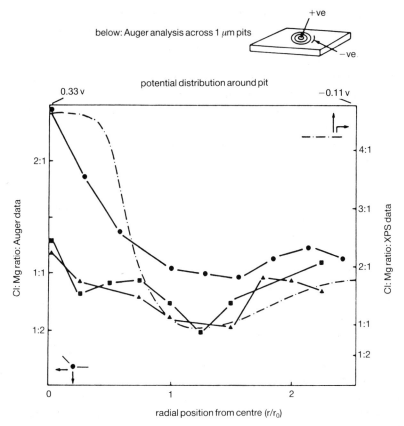

Fig. 5.6. Electrode potential around a pit. The dashed line represents the predicted potential distribution obtained from large area (XPS) analysis and consideration of pit geometry (upper and RHS axes). The data points represent Auger analyses made on very fine pits (lower and LHS axes).

## 5.4 Ceramics

In the field of ceramics it has been XPS that has proved the most useful of the two techniques, with problems of sample charging limiting the number of investigations carried out by AES. However, the last few years have seen a widening of the use of Auger spectroscopy in the fields of catalysis and mineralogy and one can be sure that this trend will continue. In this section we shall consider the role that electron spectroscopy has to play in the analysis of catalysis samples and naturally occurring minerals.

In the application of XPS to catalysis studies, there appear to be three areas of endeavour:

1. The furthering of the basic science of heterogeneous catalysis has

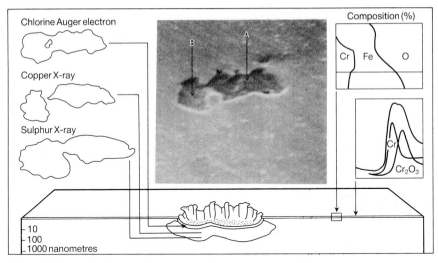

Fig. 5.7. A combination of analytical methods have been used to define the corrosion processes occurring around the inclusion group identified as sulphide containing by EDX (middle left). The passive film is characterized by Auger depth profiling and XPS, (upper right). The chlorine Auger map indicates the active area which corresponds to the pit observed in the SEM image (which is 6 $\mu$m wide).

relied greatly on the pure surface science approach, i.e. the preparation of metal or inorganic single crystals with a pre-defined crystal orientation which is then exposed to very small quantities of reactant(s) in the gas phase. In this manner, the reaction occurring on the crystal surface can be followed in a stepwise pattern, the modification of substrate or adsorbate being apparent from the electron spectrum. Such experiments are often carried out in conjunction with low energy electron diffraction (LEED) which yields information concerning surface crystallography.

2. The activity of a supported catalyst is frequently a function of the level of dispersion of the metal or oxide on the support medium. The size of such supported crystallites can sometimes be estimated from the intensity of the appropriate XPS peak or peak ratio. However, this does require assumptions regarding particle shape and the most rewarding studies appear to be those which combine XPS data with a TEM study.

3. The area in which surface analysis has made the most spectacular impact is in the identification of catalyst poisons, and other trouble-shooting investigations.

An example of the use of AES in comparing fresh and spent catalyst is presented in Fig. 5.8. The two spectra were obtained from a 0.5 per cent Pd catalyst supported on $Al_2O_3$ with chromium and molybdenum additions as promoters. Comparison of the Auger spectra from the two samples indicates that de-activation is associated with a large concentration of iron attenuating

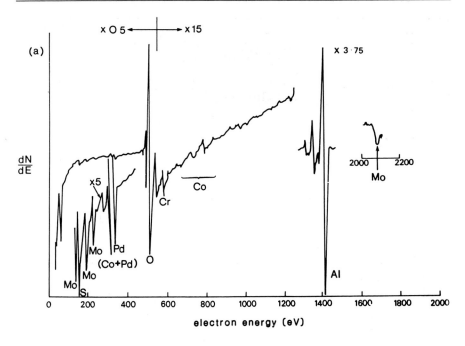

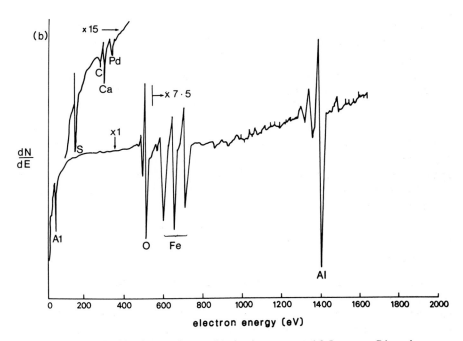

Fig. 5.8. AES of fresh (a) and spent (b) alumina supported 0.5 per cent Pd catalyst.

the Al, Pd, Cr, and Mo signals present on the clean surface. Thus the poor performance of this material could be associated with an iron contaminant, probably emanating from steel pipework or reaction vessel, masking the highly active palladium atoms as well as the promoter atoms, and greatly reducing the catalytic activity of the material.

The surface analysis of naturally occurring minerals, although straight-forward in principle, provides many problems in practice. Whilst the provision of an elemental surface analysis by XPS is straightforward, extracting the required level of chemical information can be difficult. There are two problems involved. First, electrostatic charging often means that the confidence level on the charge corrected peak position is greater than the spectral window containing the chemical information! Secondly, the peaks often have a very poor photoelectron cross-section (e.g. Si 2p, and Al 2p). The means by which these problems have been overcome are closely related to each other and have been pioneered by the Group at the University of Surrey.

The problem of sample charging may be overcome by reporting the separation of two peaks rather than the absolute binding energy of a photoelectron transition. In some elements the Auger chemical shift is equal to and sometimes greater than the XPS chemical shift, and the potential exists for extracting chemical information via the Auger parameter ($\alpha$) defined in Chapter 3, as:

$$\alpha = E_K + E_B - h\nu$$

where $E_B$ is the binding energy of the XPS peak (e.g. 1s or 2p) and $E_K$ is the kinetic energy of the attendant Auger peak (e.g. $KL_{2,3}L_{2,3}$). In the case of aluminium and silicon the $KL_{2,3}L_{2,3}$ Auger transition is not directly accessible although the Bremsstrahlung radiation from a conventional X-ray gun is able to eject sufficient Si 1s electrons to produce a measurable $SiKL_{2,3}L_{2,3}$ peak. The silicon Auger parameter calculated in this way is independent of sample charging but strongly dependent on both molecular and crystalline structure.

There still exists the problem of poor photoelectron cross-section of the Al 2p and Si 2p levels and the only way to overcome this is to use a higher energy X-ray anode able to excite 1s core levels of these elements. Various possibilities exist including $ZrL\alpha$ and $TiK\alpha$, but the one which represents the best combination of sensitivity and resolution is monochromated $AgL\alpha$. Using such a source it is possible to record 1s-$KLL$ Auger parameters; a spectrum obtained with a high energy source of this type is shown in Fig. 5.9.

One final application of XPS in the analysis of ceramics is that of photoelectron forward scattering, also referred to as X-ray photoelectron diffraction (XPD). If the specimen under consideration is a single crystal, photoelectrons emitted from sub-surface planes will be scattered by surrounding atoms giving rise to an angular distribution of the emitting site

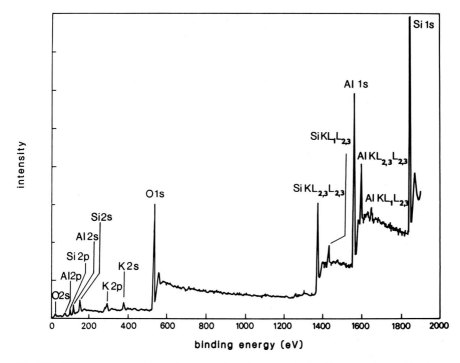

Fig. 5.9. XPS survey spectrum of muscovite mica recorded with monochromatic AgLα radiation.

concerned. This type of angular modulation is quite different to that utilized in the non-destructive depth profiling of materials described in Chapter 4. Indeed for successful forward scattering, the crystal surface should be extremely clean, preferably prepared *in vacuo* by, for instance, cleaving. This procedure has been used with a high degree of success for the investigation of complex minerals, and also semiconductors such as gallium arsenide. It would seem that such a method offers much promise as a way of ensuring the epitaxial orientation of very thin layers, the data of Fig. 5.10 shows such XPD patterns for a (100) GaAs substrate and for a thin ZnSe layer grown on it. The coincidence of the two patterns indicates the same crystal orientation in substrate and overlayer.

## 5.5 Microelectronics and semiconductor materials

Although XPS has been invaluable in determining valence states of intermediate compounds and identifying oxide overlayers, in microelectronics investigations it has been Auger spectroscopy in its depth profiling and imaging form (SAM) that has been the most widely used technique. Indeed, some of the earliest examples of applied electron spectroscopy came from

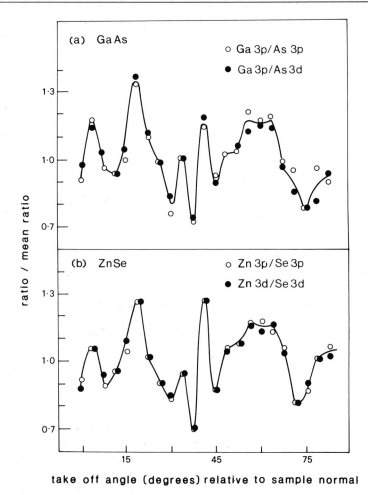

Fig. 5.10. Photoelectron forward scattering (XPD) from (a) GaAs single crystal surface, and (b) similar crystal in the same orientation with a thin film of ZnSe. The coincidence of the two sets of angular data indicate that it is an epitaxial deposit.

this area of research and it has continued to develop rapidly. A review paper from several years ago cited in excess of four hundred references! Against such a background it is impossible to do more than cite a few representative examples.

The use of Auger spectroscopy for compositional depth profiling, sometimes referred to as thin film analysis (TFA) is widely used to determine the thickness of the thin overlayers, and the effect that process parameters have on features such as oxide thickness or the level of elemental interdiffusion. The data of Fig. 5.13 illustrates the effect of raising the substrate temperature on the diffusion within a layer of $W_xSi_y$ produced by chemical vapour deposition on a silicon on insulator (SOI) substrate. The low temperature deposition shows three well defined layers, $W_xSi_y$, polycrystalline silicon, and single crystal silicon separated by discrete layers of $SiO_2$, (Fig. 5.11(a)). At the higher temperature (Fig. 5.11(b)), a great deal of oxygen diffusion within the outer layer has occurred, the stoichiometry of this layer changing with depth as the concentration of oxygen increases. This is undesirable from the

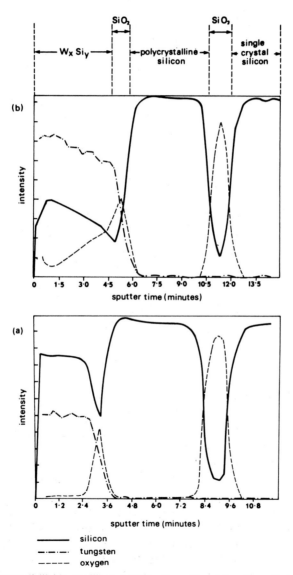

Fig. 5.11. Tungsten disilicide on silicon AES depth profiles, (a) as deposited, and (b) following heat treatment.

end users point of view as the physical and/or electrical properties of importance and gradually changing throughout the layer, unlike the example of Fig. 5.11(a) which shows a remarkable consistency through this layer, which quantitative analysis was able to identify as tungsten disilicide.

One method in which an SOI structure can be obtained is by implanting very large doses of high energy oxygen ions into a single crystal substrate. Because of their energy the oxygen ions aggregate below the surface forming a buried $SiO_2$ layer within the silicon, the extent of the buried oxide layer can be determined by Auger spectroscopy in the manner described above. The chemical shift on the silicon Auger peaks is also able to differentiate between silicon as $Si^0$ or $Si^{4+}$. However, to understand the interfacial chemistry of the $Si/SiO_2$ system fully, it is necessary to use XPS in conjunction with ion sputtering. Figure 5.12 shows a series of X-ray induced spectra from a depth profile through an interface of this type; the (Bremsstrahlung-induced) Auger peaks show a gradual change from elemental silicon to the oxide, but careful examination of the XPS spectra indicates an intermediate compound, of uncertain stoichiometry $SiO_x$. The inset of Fig. 5.12 identifies this component quite clearly, and curve fitting of all the data shows it to, appear before the main oxide peak. Supplementary transmission electron microscopy studies indicate that the actual morphology of this system to be a very fine grained oxide dispersion within a silicon matrix, the interface between the two probably having the characteristic $SiO_x$ composition.

The ability of AES to distinguish between silicon and its oxide has proved invaluable in the scanning Auger microscopy of silicon devices, where individual maps of the two phases can be produced, allowing the physical dimensions of the oxide surface features to be easily determined. Such microscopy is proving very popular within the microelectronics industry as the spatial resolution (as well as the chemical resolution) is better than that attainable in electron probe microanalysis.

## 5.6 Polymeric materials

Since its inception as a commercial technique some two decades ago, XPS has been used widely as a technique for the surface chemical analysis of polymers. Its acceptance in this role is mainly due to the group at the University of Durham, who were responsible for much of the careful characterization work in the early days of its development.

As all organic polymers contain substantial quantities of carbon, it is the chemical shift of the carbon 1s electrons which predominates in the interpretation of XPS data from these materials. Figure 5.13 shows the C1s spectrum of poly-hydroxybutyrate together with the structural formula. In order to achieve satisfactory peak synthesis of the experimental spectrum it is necessary to use four singlets. These peaks correspond to aliphatic carbon at a binding energy of 285 eV (a useful internal standard in polymer

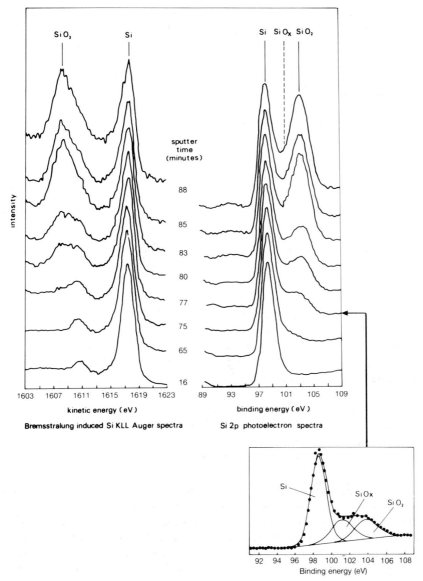

Fig. 5.12. XPS/AIB of SOI structure, Si2p and Si*KLL* showing SiO$_x$. The inset shows peak fitting of the Si$^0$, SiO$_x$, and SiO$_2$ components.

analysis), carboxyl carbon at a separation of approximately 4.2 eV from the C–C/C–H peak, and ether-like carbon at a distance of about 1.8 eV. The final component at a separation of 0.7–0.8 eV is due to a secondary chemical shift, that is the effect of the carboxyl group on the unsubstituted carbon atom in the C–CO$_2$R structure. Such secondary shifts (also known as nearest

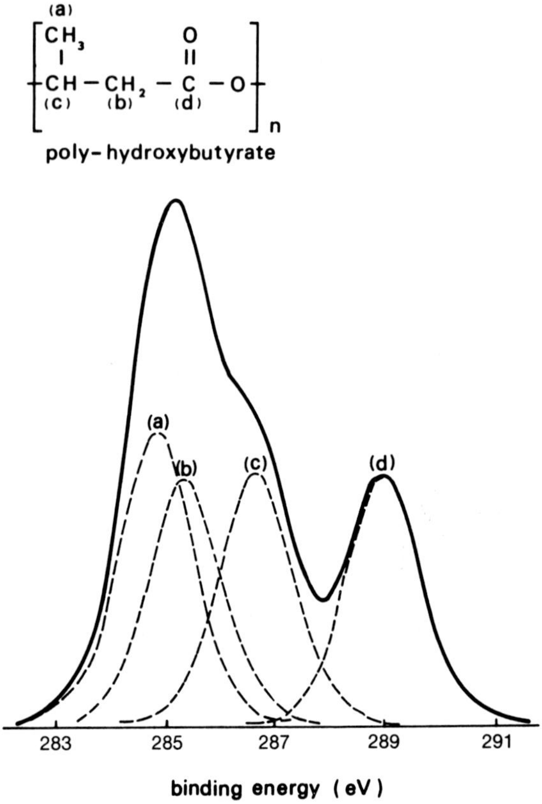

Fig. 5.13. Curve synthesis of C1s spectrum of poly-hydroxybutyrate, relating the chemical shift on the C1s electrons to the polymer structure.

neighbour effects) have only been reported in the literature relatively recently, and it is clear that their identification results from improvements in peak fitting methods. This is now invariably carried out by computer methods but much of the early work was achieved using much less sophisticated techniques. In such instances, the secondary shift is accounted for by merely making the methyl carbon peak slightly wider and more intense. Unless the carboxyl peak is fairly strong, the secondary shift is easily lost in the vagaries of the peak fitting exercise. Inspection of the structure of the poly-hydroxybutyrate molecule shows that there are equal amounts of each type of carbon in the repeat unit. The three chemically shifted carbon components of Fig. 5.13 are of equal intensity but the aliphatic C–C/C–H component is slightly more intense, this may be the result of surface segregation and/or orientation effects, or more likely, a small amount of adventitious contamination on the specimen.

By careful use of XPS, it is possible to differentiate between aliphatic and aromatic carbons. The spectrum of Fig. 5.14 is taken from polystyrene and

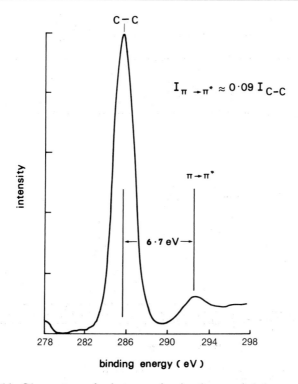

Fig. 5.14. C1s spectrum of polystyrene showing the $\pi \rightarrow \pi^*$ shake-up satellite.

the shake up satellite resulting from the $\pi \rightarrow \pi^*$ transition in the phenyl ring which accompanies photoemission can be seen as a discrete feature some 6.7 eV from the main peak. The intensity of the satellite as a function of the main photoelectron peak remains constant at around 10 per cent although slight changes occur depending on the structure of the polymer involved. This feature provides a quantitative way in which the surface concentration of phenyl groups following a particular treatment method may be estimated. It also provides a means of estimating surface modification brought about by ring opening reactions.

In all but very special cases, depth profiling of polymers by ion sputtering is impractical because of gross sample degradation and the usual methods adopted are angle resolved XPS or multi-photon investigations. An example of the use of angle resolved XPS, Fig. 5.15, shows the variation in near-surface composition of a segmented polyurethane film, the surface enhancement of oxygen and carbon bonded to oxygen and nitrogen (C–O and C–N) is seen. This depth profile should be viewed as a linear 'pie-chart', the area of each field representing the relative abundance of each species as a function of depth. The alternative method of depth-profiling, changing the energy of the X-ray photons, is potentially a very elegant way in which to probe

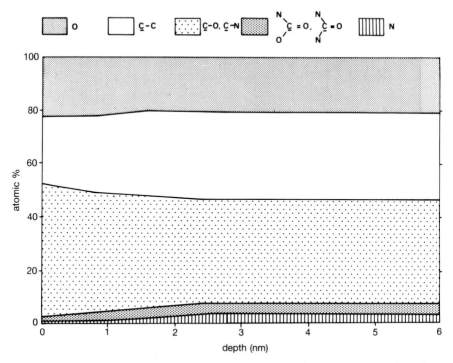

Fig. 5.15. Reconstructed depth profile from angle resolved XPS data of a cast polyurethane film, showing surface enhancement of oxygen.

different depths although in practice it is rather restrictive and most spectroscopists will be limited to a combination of AlK$\alpha$ and MgK$\alpha$.

There are certain instances where the concentration of particular functional groups at a polymer surface are very low and curve fitting of the carbon 1s spectrum can become a rather uncertain process. In these instances, a method known as chemical derivatization can then be used. In essence the surface functional groups are reacted with a liquid or gas phase reagent which tags them with an atom or ion which is readily determined by XPS, the concentration of the characteristic element is then directly proportional to the concentration of the functional group involved. Many such derivatization reactions exist using both organic and inorganic reagents, the most successful being those where the chemical tag has a high cross-section in XPS such as barium, thallium, or silver.

The vast majority of polymer science investigations making use of XPS report the changes that have occurred as a result of surface modification, either by process treatment (to improve surface properties such as wettability), or naturally occurring phenomena (such as the weathering of paint films).

## 5.7 Adhesion science

There are three distinct areas in which surface analysis has made major contributions in the science and technology of adhesion; the analysis of surfaces prior to application of the adherate, and the subsequent correlation of adhesion with surface cleanliness; the investigation of the substrate to polymer bond; and the exact definition of the locus of failure following bond failure. Each of these areas will now be considered in turn.

The cleanliness of a metal substrate prior to its contact with a polymer adhesive or coating is readily assessed by XPS or AES, and although gross levels of contamination such as oils from drawing or temporary corrosion protection can easily be detected by other methods, it is only surface sensitive techniques which can assess the efficacy of the cleaning method. As an example, the spectra of Fig. 5.16 are taken from steel sheet that has been

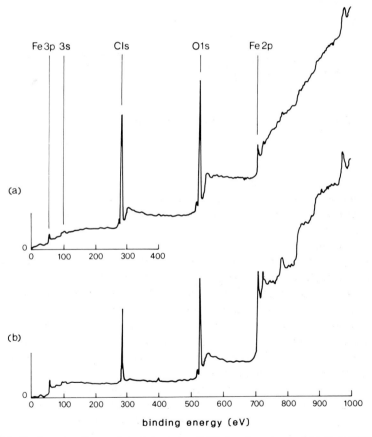

Fig. 5.16. Surface cleanliness of steel sheet. An alkali cleaning process (a) leaves substantially more contamination than an abrasive cleaning process (b).

prepared by alkaline cleaning (Fig. 5.16(a)) and emery abrasion (Fig. 5.16(b)), the level of carbonaceous contamination is considerably higher on the solution cleaned surface indicating that abrasive cleaning produces the better quality surface from a chemical point of view. Studies some ten years ago by the American automobile industry unequivocally established high levels of surface carbon as a contributory factor to poor durability of painted steel from certain manufacturers. Another frequently cited example is that of aluminium–magnesium alloys, which, if heat treated incorrectly, develop a friable surface film of magnesium oxide. Adhesion of a paint film or adhesive to this layer is very poor and failure occurs rapidly. XPS or AES is able to identify such a layer and these techniques can be used in a diagnostic manner prior to bonding or painting to ensure that adhesion will be of the required standard. Contamination may also arise from a variety of external sources, sub-monolayer coverage of aluminium surfaces by fluoro-carbons has been shown to produce a drastic reduction in bond strength.

Once the bond has been made, the task of examining the interfacial chemistry is extremely difficult. Various methods of approaching the interface have been developed, involving the removal of the polymer in a suitable solvent, or the dissolution of the iron substrate in a methanolic iodine solution followed by sputter depth profiling through the oxide towards the interface, but both suffer from their own particular problems. The last year or so has indicated that careful ultramicrotomy followed by STEM used in conjunction with windowless EDX, EELS, or electron diffraction may be more useful where an interphase has developed, but the analysis of the interface on an atomic scale by these methods is still some way off.

The analysis of a failed interface is routinely carried out by electron spectroscopy and the definition of adhesive or cohesive failure, on an atomic scale, has become a straightforward matter for those working in the field. The spectra of Fig. 5.17 are from the substrate (steel) side of two markedly different failures. Figure 5.17(a) shows a true interfacial failure for a polymer coating peeled from a steel surface, while Fig. 5.17(b) illustrates the case of a cohesive failure where a thin layer of polymer remains on the substrate surface. The differences in the spectra are striking. The organic residue, in the case of the cohesive failure increasing the intensity of the carbon 1s peak and attenuating the iron 2p peaks (at about 710 eV) completely. Both light and electron microscopy identified the latter failure as adhesive because the overlayer of polymer was too thin to have been detected by such methods.

The use of XPS and AES in the type of studies described above has been carried out by a few groups for the last decade. A more recent development has been their use in studies of composite materials. XPS has been used for some time to assess the surface acidity of carbon fibres and the level of sizing, but work is now being undertaken to study composite fracture surfaces by these methods. Already, both XPS and AES have been used to study the interfacial region in metal matrix composites, and it appears that, in some

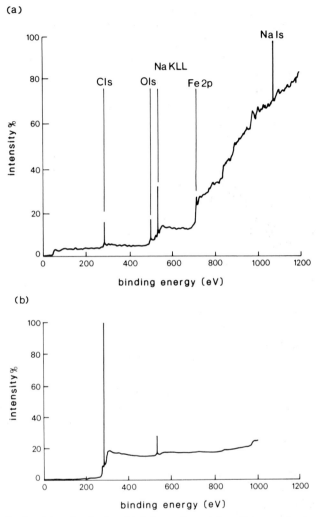

Fig. 5.17. Adhesive (a) and cohesive (b) failure of a polybutadiene coating applied to mild steel.

cases, minor elements from the alloy matrix may segregate to the (ceramic) fibre surface. Investigations in these areas are ongoing and it is clear that a degree of technique development is needed before the full potential of XPS and AES is realized in these fields.

　While the examples cited in this chapter have dealt exclusively with the techniques described earlier in this book, it would be a very narrow minded scientist who did not make full use of the plethora of advanced analytical techniques that are becoming available. In the next chapter the more common ones are described and comparisons drawn with XPS and AES.

# 6 Comparison of XPS and AES with other analytical techniques

A recent compilation of physical examination and analytical techniques identified almost 150 methods which could be used for materials analysis. The set of acronyms assigned to these methods is now vast and inevitably, confusion has arisen for the surface scientist. For example, SAM stands for scanning Auger microscopy but an equally acceptable meaning is scanning acoustic microscopy. The majority of these techniques are specialist methods requiring careful specimen preparation and experimentation, others are applicable to a fairly limited portion of the periodic table or accept specimens in only one particular form. The aim of this section is to compare the subjects of this text with other analytical methods that are available within research institutes and academia. We shall exclude those processes which yield structural information, such as X-ray and electron diffraction; also excluded are the various vibrational spectroscopies that yield essentially molecular rather than elemental analyses, such as infra-red and Raman spectroscopy. The classification of analysis methods may be carried out in several ways but, for the time being, we shall consider them in terms of primary (incident) and secondary (emitted) radiations, according to Table 6.1, which lists the nine, of the many possible methods, that we shall consider.

The acronyms used have the following meanings:

EDX     energy dispersive X-ray analysis
EELS    electron energy loss spectroscopy
ISS      ion scattering spectroscopy
LAMMS laser ablation microprobe mass spectrometry
RBS     Rutherford backscattering spectrometry
SIMS    secondary ion mass spectrometry
WEDX  windowless energy dispersive X-ray analysis

## 6.1 X-ray analysis in the electron microscope

The addition of an X-ray analyser (either energy or wavelength dispersive, EDX or WDX) to a scanning electron microscope provides an electron probe microanalysis facility widely used in all branches of research and development. This provides a very flexible means of microanalysis and, in the conventional EDX mode, elements from sodium onwards can be detected; with a WDX spectrometer, lighter elements down to oxygen are accessible.

Table 6.1. *Features of various analytical methods discussed in the text*

| | Incident radiation | Emitted radiation | Property monitored | Elements detectable | Depth of analysis | Spatial resolution | Information level E = elemental C = chemical | Quanti-fication # | Applica-bility to inorganics # | Applica-bility to organics # |
|---|---|---|---|---|---|---|---|---|---|---|
| AES | e⁻ | e⁻ | energy | Li on | 3 nm | 50 nm | E(C) | ✓ | 0 | X |
| EDX | e⁻ | X-rays | energy | Na on | 1 µm | 1 µm | E | ✓ | 0$ | X$ |
| EELS | e⁻ | e⁻ | energy | Li on | depends on foil thickness | 10 nm | E | 0 | ✓ | X |
| ISS | ions | ions | energy | Li on | outer atom layer | 1 mm | E | 0 | ✓ | 0 |
| LAMMS | laser | ions | mass | all | 0.5 µm | 1 µm | E, C | X | ✓ | ✓ |
| RBS | ions | ions | energy | Li on | 1 µm | 1 mm | E | X | ✓ | ✓* |
| SIMS (static) | ions | ions | mass | all | 1.5 nm | 1 mm | C(E) | X | ✓ | ✓ |
| SIMS (dynamic) | ions | ions | mass | all | see text | 50 µm | C(E) | 0 | ✓ | X |
| SIMS (imaging) | ions | ions | mass | all | see text | 200 nm | C(E) | X | 0 | 0 |
| WEDX | e⁻ | X-rays | energy | B on | 1 µm | 1 µm STD 10 mm² small area: 100 µm imaging XPS: 10 µm | E | ✓ | 0$ | X$ |
| XPS | X-rays | e⁻ | energy | He on | 3 nm | | E,C | ✓ | ✓ | ✓ |

# ✓ = good, 0 = reasonable, X = poor.
$ without conductive coating.
*cryo-stage required.

The great advance in recent years is the advent of the windowless EDX (WEDX) detector which extends the range to much lighter elements, from boron onwards, but the vacuum requirements are more stringent to prevent icing up of the detector. However, although it has now become possible to undertake light element analysis by the consideration of emitted characteristic X-rays, such analyses are essentially probing bulk composition. The 'interaction volume' of the electron beam with the sample determines both the lateral resolution (in the analytical mode) and the depth of analysis; this is a function of primary beam energy but will invariably be of the order of a micrometre. By reducing beam energy, the depth of analysis may be reduced to as little as 500 nm but there is a limit to this approach, for although an electron beam of 1 keV will only have a small penetration depth it will not excite X-rays of analytical use in the conventional sense. The characteristic electron inelastic mean free path ($\lambda$) for electrons of analytical use in electron spectroscopy is compared with that for the higher energies used as primary radiation in the electron microscope in Fig. 6.1. Thus the analysis depth in X-ray analysis is determined by the energy of the primary radiation (the electron beam) whereas in electron spectroscopy it is the energy of the secondary radiation (emitted electrons) that controls this parameter.

In the TEM, the use of a thin foil specimen immediately defines the depth of analysis as the pear shaped interaction volume is abruptly truncated. This dramatically improves spatial resolution, although there will be some degradation of spatial resolution as a result of electron beam interaction with the sample. A WEDX spectrum of pure boron (with a thin oxide film present), obtained in a TEM, is shown in Fig. 6.2(a), the BK$\alpha$ line is very clear in the

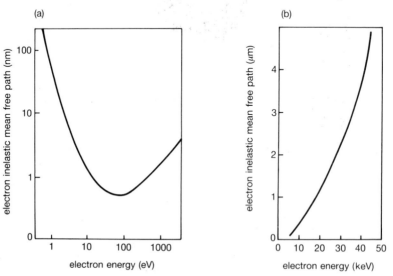

Fig. 6.1. Electron mean free path of energies used in electron spectroscopy (a), and of the primary beam energies used in electron microscopy (b).

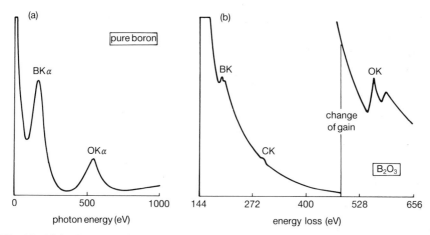

Fig. 6.2. Light element analysis in the TEM by (a) WEDX and (b) EELS of boron and $B_2O_3$.

spectrum. In the scanning Auger microscope the analysis depth is determined by the energy of the outgoing electrons as described in Chapter 1, and the spatial resolution depends on the size of the electron probe. Thus the spatial resolution attainable with SAM can be an order of magnitude better than that recorded with SEM/EDX. With the development of sub-micron features in microelectronics SAM is finding a new use; as a high resolution chemical imaging facility, the emphasis no longer being on the need for a *surface* analysis.

Although X-rays do show a small chemical shift with oxidation state this feature is not employed in analytical X-ray analysis of the type used in electron microscopes thus EDX only provides elemental information unlike the additional chemical information provided by XPS and occasionally by AES.

The addition of an EDX facility to a surface analysis spectrometer is worthy of consideration. In the SAM mode it is possible to acquire both Auger (surface) and X-ray (bulk) chemical maps of the specimen. In conjunction with XPS, an X-ray detector is able to provide good quality fluorescence spectra (XRF) to within about 2 kV of the X-ray source operating potential. As this will usually be 12–15 kV, X-ray spectra up to 10–13 kV can be obtained. This method is particularly useful for insulating specimens not amenable to analysis by AES/EDX such as paint films or some catalysts. XRF is also useful in XPS depth profiling where a global analysis of specimen chemistry can be achieved before segregation or interfacial effects are studied in detail.

## 6.2 Electron analysis in the electron microscope

As well as the possibility of X-ray analysis in the TEM and STEM it is possible to analyse the energy of the transmitted beam which is the basis of electron energy loss spectroscopy (EELS). As an electron beam passes through an electron transparent specimen it is able to eject electrons whose binding energies are less than that of the primary beam energy. By recording the depletion of the primary beam energy with an electron spectrometer positioned below the specimen an energy loss spectrum can be obtained. Such a spectrum will have characteristic edges at energy loss values equivalent to the core electron's binding energy. As any electron of energy loss greater than the binding energy can cause ionization, the edge has an abrupt start at the binding energy but tails off as the energy loss increases and could decay until the primary beam energy is reached (i.e. almost total loss of energy). In practice, the cross section falls as $E^{-4}$ and the resultant inner shell edge resembles a triangle, as can be seen in the EELS spectrum of Fig. 6.2(b). The energies of these characteristic edges are close to the binding energies used in XPS; e.g. carbon $K$ edge is 284 eV, magnesium $K$ edge is 1305 eV, copper $L_{2,3}$ edge (convolution of Cu $2p_{3/2}$ and Cu $2p_{1/2}$) is 941 eV. The major problems of quantifying EELS spectra are accurately stripping off the background and deconvoluting overlapping or adjacent peaks. Once these difficulties have been overcome completely the outlook appears very promising. In addition, EELS mapping can be carried out in the STEM, producing chemical images of very light elements. Once again, EELS does not generally have the chemical specificity of emitted electron spectroscopy, although some elements do show a chemical shift analogous to that of XPS. Specimen preparation can be complex and/or tedious. Its one great advantage (shared by EDX or WEDX in the STEM) is that of spatial resolution which is determined by the resolution of the microscope plus the 'beam spreading' contribution. The depth of analysis is clearly defined by the foil thickness which will be of the order of tens of nanometres. The very important role that EELS has to play is in the identification of very fine bulk features containing light elements, such as oxide, carbide, and nitride precipitates in steels.

## 6.3 Mass spectrometry for surface analysis

In parallel with the growth and development of XPS and AES over the last two decades, the technique of secondary ion mass spectrometry (SIMS) has also been evolving. The basic principle of SIMS is the bombardment of a materials surface with a beam of energetic ions and the subsequent mass analysis of the sputtered ions and cluster ions. The great strength of this method is that all elements including hydrogen can be detected and as the analysis is based on the mass separation of the secondary particles it has isotopic and molecular specificity.

There are three modes of operation in SIMS.

1. Static SIMS (SSIMS) were the low ion flux ensures that the amount of material removed during analysis is sub-monolayer, this version is invariably carried out with a defocused ion beam;

2. Dynamic SIMS, in which sputtering proceeds at a fairly rapid speed (as in AES/AIB compositional depth profiling) and only those mass fragments of interest are monitored and plotted as a function of sputter time. Examples of these two modes of operation are shown in Figure 6.3. Recently, SSIMS has been applied to polymers and other insulating specimens with a great deal of success, in particular the use of xenon or argon atoms (rather than ions) has been shown to reduce charging effects and sample damage usually associated with these materials. This technique, known as fast atom bombardment mass spectrometry (FABMS) has greatly extended the use of mass spectrometry for surface analysis.

3. Imaging SIMS, which can be achieved in two different ways, relies on either the resolution of the ion beam itself or the ability of the spectrometer to retain spatial information as the secondary ions pass through a magnetic analyser. In the former approach a fine, sub-micron, beam of perhaps gallium ions is rastered across the surface and a particular fragment is monitored and used to build up a chemical map of the surface in a sequential manner, as in EPMA or SAM. This mode of acquisition is referred to as the scanning ion microprobe. The alternative, more widely practised, method is known as ion microscopy, in which the entire field of view is imaged simultaneously in the chosen ion species. In this type of imaging SIMS, the spatial resolution attainable is defined by the size of the apertures in the ion optics and, at the ultimate resolution (0.5 $\mu$m), by the optical aberrations of the spectrometer itself. The spatial resolution of this type of instrument is independent of primary beam size and in this respect it can be considered to be analogous to the optical microscope.

A SIMS analysis is also extremely surface sensitive with an analysis depth in the static SIMS mode of around 1.5 nm. With dynamic SIMS, the material is being sputtered at such a rate that the actual surface is constantly being eroded and the term depth of analysis  becomes less important; it is now the depth resolution that is the parameter by which the technique is judged. The detection limit is probably one of the best obtainable with the methods discussed in this chapter, being of the order of p.p.b. in favourable cases. Quantification can be carried out quite accurately by comparison of the specimen being examined with standards of very close composition, but for routine analysis of unknown specimens, quantification will not usually be attempted.

A method closely associated with SIMS, in that the mass analysis of emitted ionized atoms, molecules, and clusters is undertaken, sometimes on the same analytical system, is laser ablation microprobe mass spectrometry (LAMMS). The lateral and depth resolutions attainable with LAMMS are

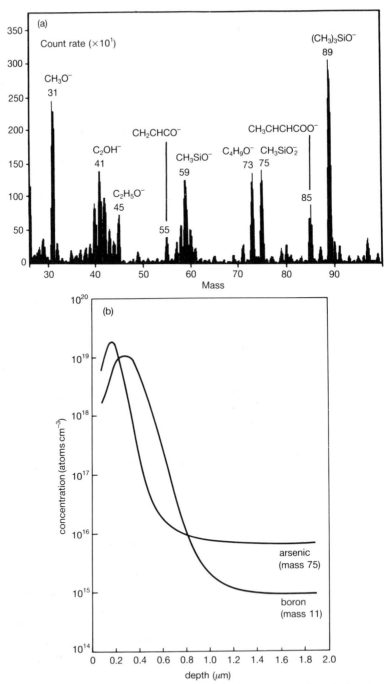

Fig. 6.3. (a) SSIMS of polymeric materials, the negative ion spectrum of an alkyl methacrylate organosiloxane co-polymer, (b) a dynamic SIMS depth profile of arsenic and boron in silicon.

dependent on both the power and the focus conditions of the pulsed laser team. A fully focused beam will provide very good spatial resolution but as the power density at the sample surface is very high and consequently the crater formed at the sample surface will be relatively deep; the so-called 'hard' laser ionization mode. For the same power output a fully defocused beam will create a crater which has a larger diameter but is much shallower; the 'soft' laser desorption mode, (0.1 $\mu$m is probably the best 'sampling depth' achievable with LAMMS at present). It is interesting to note that this technique was originally developed as a bulk microprobe instrument complementary to EDX and giving elemental and isotopic information down to hydrogen. It is available in both the reflection mode analogous to SEM/EDX and transmission configuration, the TEM/EDX equivalent. Although the mass spectrum obtained by LAMMS does not give an *exclusive* surface analysis in the same way that SSIMS, XPS, or AES does, it will often provide valuable information concerning the surface phases. This is because in the ablation process the volatilized material of the crater will necessarily include that at the very surface, the high sensitivity of mass analysis techniques such as SIMS and LAMMS ensure that any unsuspected elements present at the surface are clearly defined in the resultant mass spectrum. The major advantages of LAMMS at the present time are its ability to act as a light element/isotopic specific microprobe, and the rapidity with which it can profile thin films; the depth of each laser shot can be matched to film thickness by varying the operating conditions from 'soft' to 'hard'. Unlike SEM and SIMS insulating samples can be analysed without any special sample preparation or charge compensation being required. The main problem is quantification of the resultant spectrum, although semi-quantitative data can be achieved quite successfully using standards of similar composition. In the future, it seems likely that the use of very low laser powers for true surface studies will see the development of LAMMS into a valuable analytical technique to complement SSIMS.

## 6.4 Other ion beam techniques

In this section we shall consider briefly the scattering of incident ions by a solid sample. This forms the basis of two, markedly different, analytical techniques. In the case of low energy ions (0.2–3 keV) the method is known as ion scattering spectroscopy (ISS) or more correctly as low energy ion scattering spectroscopy (LEIS), and the resultant analysis is extremely surface sensitive. When higher energies (1–5 MeV) are used the primary ions are back-scattered from the target atoms and the resultant spectrum can be interpreted not only in terms of elements present but also as a depth profile to a depth of about 1 $\mu$m. High energy ion scattering spectrometry (HEIS) is generally referred to as Rutherford backscattering spectrometry (RBS).

ISS is often performed on a surface analysis system equipped for electron spectroscopy, the electron energy analyser being modified to facilitate the detection of either electrons (for XPS or AES) or the scattered ions. The specimen is bombarded by a monoenergetic ion beam (usually $He^+$, $Ne^+$, or $Ar^+$) of energy $E_0$, the ions are scattered elastically from the atoms of the outermost layer of the solid and their energy $(E)$ is measured by the energy analyser. For a binary collision and a scattering angle of 90° there is a direct relationship between the kinetic energy of the scattered ion and the mass of the surface atom with which it has collided:—

$$E/E_0 = (M_2 - M_1)/(M_2 + M_1)$$

where $M_1$ = mass of the incident ion, $M_2$ = mass of surface atom. The ion scattered spectrum takes the form of the intensity of the scattered primary beam as a function of its energy normalized to the primary beam energy $(E/E_0)$, as shown in Fig. 6.4. One of the main difficulties of employing ISS stems from the very surface sensitive nature of the analysis and surface contamination may become a serious problem. There are various ways of circumventing this involving *in situ* sample preparation by heating, ion sputtering, or cleaving.

In the case of RBS the energy of the primary ion beam is much higher and the experimental set-up will include some form of accelerator to provide ions of sufficient energy (around 2 MeV). The RBS experiment consists of bombarding the surface of the specimen with these high energy ions (usually

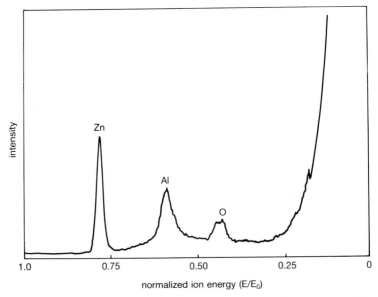

Fig. 6.4. ISS spectrum from a galvanized steel surface showing the presence of aluminium added to the bath to enhance the appearance.

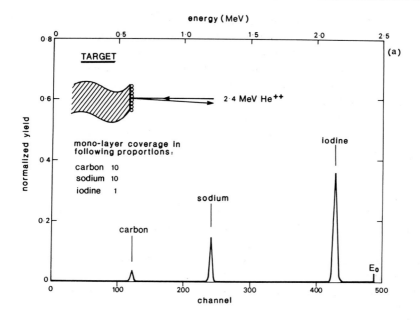

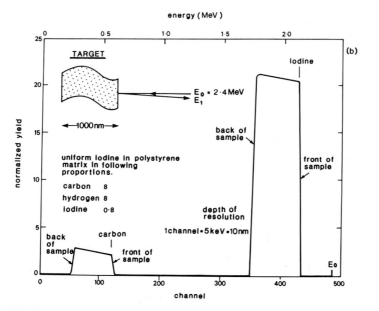

Fig. 6.5. Simulated RBS spectra illustrating, (a) the cross-section dependence on atomic number (varies with $Z^2$), and (b) depth profiling capabilities through a 1 $\mu$m film.

$He^{++}$) and simultaneously measuring the energy of the backscattered ion ($E$). This energy is given (for a scattering angle of 90°) by:

$$E = E_0(M_2 - M_1)/(M_2 + M_1)^2$$

Thus the primary ions that interact with light elements become backscattered ions of low energy, and those primary ions that interact with higher masses are scattered with a higher energy. However, scattering takes place not only at the surface, as in the case of ISS, but many atomic layers into the sample and the ions undergo attenuation on their way both into and out of the specimen. Consequently, the RBS spectrum contains not only information concerning the identity of the atoms within the interaction volume but also their position in relation to the free surface. Because of this convolution of elemental and depth information in the final spectrum, care must be taken in interpretation, and it is usual to restrict its application to systems with only a few known elements present. A simulation programme will then be employed to reconstruct the depth distribution of the elements in the specimen accurately. RBS is widely used in the analysis of dopants and thin films used in the microelectronics industry. Quantification can be performed accurately from first principles methods using experimental parameters and scattering cross sections; detection limits are typically in the 100 p.p.m. range, but can vary very widely depending on the matrix material. The calculated RBS spectrum of Fig. 6.5(a) illustrates the variation in cross section with increasing mass for a monolayer of material (carbon 10: sodium 10: iodine 1) deposited on a substrate which does not contribute to the spectrum. The depth dependency of the RBS spectrum is seen in Fig. 6.5(b), where the specimen is a 1 $\mu$m thick deposit of material (carbon 8: hydrogen 8: iodine 0.8). In this case, the individual peaks have broadened considerably and the front and back of the sample can be identified, thus providing a nondestructive depth profile of the sample on a micron scale.

## 6.5 Concluding remarks

Of the methods discussed in this chapter, those based on conventional electron microscopes, especially EDX, are readily available in universities and research establishments, and most scientists with a need for microanalysis will have encountered them. The other methods are less widely available and general access to them, outside of corporate research laboratories and universities, can only be obtained at one of the specialist surface analysis consultancy services that are available in Europe and North America.

# Bibliography

## Chapter 1

Briggs, D. and Seah, M. P. (1990). *Practical surface analysis by Auger and X-ray photoelectron spectroscopy* (2nd edn). John Wiley and Sons Ltd, Chichester, UK.

Brundle, C. R. and Baker, A. D. *Electron spectroscopy: theory, techniques, and applications.* Academic Press Inc., New York, Vol. 1 (1977), Vol. 2 (1978), Vol. 3 (1979), Vol. 4 (1981).

Carlson, T. A. (1976). *Photoelectron and Auger spectroscopy.* Plenum, New York.

Castle, J. E. (1982). Electron spectroscopy methods. In *Analysis of high temperature materials* (ed. O. Van der Biest), pp. 141–88, Applied Science Publishers Ltd, London.

Ferguson, I. F. (1989). *Auger microprobe analysis,* Adam Hilyer Ltd, Bristol, UK.

Nefedov, V. I. (1988). *X-ray photoelectron spectroscopy of solid surfaces.* VSP BV, Utrecht, The Netherlands.

Riviere, J. C. (1982). Auger techniques in analytical chemistry: a review. *The Analyst,* **108**, 649–84.

Seah, M. P. and Dench, W. A. (1979). Quantitative electron spectroscopy of surfaces: a standard data base for electron inelastic mean free paths in solids. *Surf. Interf. Anal.,* **1**, 2–11.

Seighbahn, K. *et al.* (1967). *ESCA: atomic, molecular, and solid state structure studied by means of electron spectroscopy.* Almqvist and Wiksells, Uppsala, Sweden.

Thompson, M., Baker, M. D., Christie, A. B., and Tyson, J. F. (1985). *Auger electron spectroscopy.* John Wiley and Sons Inc., New York.

## Chapter 2

Barrie, A. (1977) Instrumentation for electron spectroscopy. In *Handbood of ultra-violet and X-ray photoelectron spectroscopy* (ed. D. Briggs), pp. 79–119, Heyden and Sons Ltd, London, UK.

Brooker, A. D. and Castle, J. E. (1986). Scanning Auger microscopy: resolution in time, energy and space. *Surf. Interf. Anal.,* **8**, 113–19.

Brundle, C. R., Roberts, M. W., Latham, D., and Yates, K. (1974). An ultrahigh vacuum electron spectrometer for surface studies. *J. Elec. Spec.,* **3**, 241–61.

Castle, J. E. and West, R. H. (1980). The utility of bremsstralung induced Auger peaks. *J. Elec. Spec.,* **18**, 355–8.

Coxon, P., Krizek, J., Humpherson, M., Wardell, I. R. M. (1990). Escascope—a new imaging photoelectron spectrometer. *J. Elec. Spec.,* **52**, 821–838.

Edgell, M. J., Paynter, R. W., and Castle, J. E. (1985). High energy XPS using a monochromated AgLα source: resolution, sensitivity and photoelectric cross sections. *J. Elec. Spec.,* **37**, 241–56.

Koenig, M. F. and Grant, J. T. (1985). Monochromator versus deconvolution for XPS studies using a Kratos ES300 system. *J. Elec. Spec.*, **36**, 213–25.

Seah, M. P., Anthony, M. T., and Dench, W. A. (1983). Characterisation of computer differentiation of AES spectra and its relation to differentiation by the modulation technique. *J. Phys. E.* **16**, 848–57.

Sherwood, P. M. A. (1983). Data analysis in X-ray photoelectron spectroscopy. In *Practical surface analysis by Auger and X-ray photoelectron spectroscopy* (ed. D. Briggs and M. P. Seah), pp. 445–75, John Wiley and Sons Ltd, Chichester, UK.

Turner, D. W., Plummer, I. R., and Porter, H. Q. (1984). Photoelectron emission: images and spectra. *J. Microscopy*, **136**, 259–77.

Wagner, C. D. and Joshi, A. (1988). The Auger parameter, its utility and advantages: a review. *J. Elec. Spec.*, **47**, 283–313.

## Chapter 3

Castle, J. E., Abu-Talib, I., and Richardson, S. A. (1985). The use of energy loss structures in XPS characterisation of surfaces. *Mat. Res. Symp. Proc.*, **48**, 471–9.

Davis, L. E., MacDonald, N. C., Palmberg, P. W., Riach, G. E., and Weber, R. E. (1978). *Handbook of Auger electron spectroscopy.* Physical Electronics Division, Perkin-Elmer Corporation, Eden Prairie, USA.

Fuggle, J. C. and Martenson, N. (1980). Core-level binding energies in metals. *J. Elec. Spec.*, **21**, 275–81.

Hall, P. M. and Morabito, J. M. (1979). Matrix effects in the quantitative Auger analysis of dilute alloys. *Surf. Sci.*, **83**, 391–405.

Hall, P. M., Morabito, J. M., and Conley, D. K. (1977). Relative sensitivity factors for Auger analysis of binary alloys. *Surf. Sci.*, **62**, 1–20.

Seah, M. P. (1979). Quantitative AES: via the energy spectrum or the differential? *Surf. Interf. Anal.*, **1**, 86–90.

Seah, M. P. (1980). The quantitative analysis of surfaces by XPS: a review. *Surf. Interf. Anal.*, **2**, 222–39.

Seah, M. P. (1983). A review of quantitative Auger electron spectroscopy. Scanning Electron Microscopy, SEM Inc., Chicago, USA, **2**, 521–36.

Wagner, C. D., Riggs, W. M., Davis, L. E., Moulder, J. E., and Muilenberg, G. E. (1979). *Handbook of X-ray photoelectron spectroscopy.* Physical Electronics Division, Perkin-Elmer Corporation, Eden Prairie, USA.

Wagner, C. D., Davis, L. E., Zeller, M. V., Taylor, J. A., Raymond, R. H., and Gale, L. H. (1981). Empirical sensitivity factors for quantitative analysis by electron spectroscopy for chemical analysis. *Surf. Interf. Anal.*, **3**, 211–25.

## Chapter 4

Hofmann, S. (1983). Depth profiling. In *Practical surface analysis by Auger and X-ray photoelectron spectroscopy*, pp. 141–79, John Wiley and Sons Ltd, Chichester, UK.

Kelly, R. (1985). On the role of gibbsian segregation in causing preferential sputtering. *Surf. Interf. Anal.*, **7**, 1–7.

Lea, C. and Seah, M. P. (1981). Optimized depth resolution in ion-sputtered and lapped compositional profiles with Auger electron spectroscopy. *Thin Solid Films*, **75**, 67–86.

Paynter, R. W. (1981). Modification of the Beer–Lambert equation for application to concentration gradients. *Surf. Interf. Anal.*, **3**, 186–7.

Seah, M. P. (1981). Pure element sputtering yields using 500–1000 eV argon ions. *Thin Solid Films*, **81**, 279–87.

Seah, M. P. and Hunt, C. P. (1983). The depth dependence of the depth resolution in composition-depth profiling with Auger electron spectroscopy. *Surf. Interf. Anal.*, **5**, 33–7.

Walls, J. M., Hall, D. D., and Sykes, D. E. (1979). Compositional-depth profiling and interface analysis of surface coatings using ball-cratering and the scanning Auger microprobe. *Surf. Interf. Anal.*, **1**, 204–10.

Yih, R. S. and Ratner, B. D. (1987). A comparison of two angular dependent ESCA algorithms useful for constructing depth profiles of surfaces. *J. Elec. Spec.*, **43**, 61–82.

## Chapter 5

Baer, D. R. (1984). Solving corrosion problems with surface analysis. *Appl. Surf. Sci.*, **19**, 382–96.

Bhasin, M. M. (1975). Auger spectroscopic analysis of the poisoning of a commercial palladium–alumina hydrogenation catalyst. *J. Catalysis*, **38**, 218–22.

Briggs, D. (1983). Applications of XPS in polymer technology. In *Practical surface analysis by Auger and X-ray photoelectron spectroscopy* (ed. D. Briggs and M. P. Seah), pp. 359–96. John Wiley and Sons Ltd, Chichester, UK.

Brinen, J. S., Graham, S. W., Hammond, J. S., and Paul, D. F. (1984). Characterization of fresh and spent HDS catalysts by Auger and X-ray photoelectron spectroscopies. *Surf. Interf. Anal.*, **6**, 68–73.

Castle, J. E. (1986). The role of electron spectroscopy in corrosion science. *Surf. Interf. Anal.*, **9**, 345–56.

Castle, J. E. and Watts, J. F. (1988). The study of interfaces in composite materials by surface analytical techniques. In *Interfaces in polymer, ceramic, and metal matrix composites* (ed. H. Ishida), pp. 57–71. Elsevier Science Publishing Co. Inc.

Dilks, A. (1981). X-ray photoelectron spectroscopy for the investigation of polymeric materials. In *Electron spectroscopy: theory, techniques and applications* (ed. C. R. Brundle and A. D. Baker), Vol. 4, pp. 277–359, Academic Press Inc., New York.

Evans, S. and Scott, M. D. (1981). Chemical and structural characterization of epitaxial compound semiconductor layers using X-ray photoelectron diffraction. *Surf. Interf. Anal.*, **3**, 269–71.

Fakes, D. W., Newton, J. M., Watts, J. F., and Edgell, M. J. (1987). Surface modification of a contact lens co-polymer by plasma-discharge treatments. *Surf. Interf. Anal.*, **10**, 416–23.

Heckingbottom, R. (1986). Perspectives in surface and interface analysis for electronic devices and circuits. *Surf. Interf. Anal.*, **9**, 265–74.

Holloway, P. H. and McGuire, G. E. (1980). Characterization of electronic devices and materials by surface sensitive analytical techniques. *Appl. Surf. Sci.*, **4**, 410–44.

Muhammad, Azali bin (1982). The effect of carbides on temper embrittlement of commercial 2.25Cr–1.0Mo steel. Unpublished PhD Thesis, University of Surrey.

Paynter, R. W. and Ratner, B. D. (1985). The study of interfacial proteins and biomolecules by X-ray photoelectron spectroscopy. In *Surface and interfacial aspects of biomedical polymers* (ed. J. D. Andrade), pp. 189–216. Plenum Press, New York.

Pijpers, A. P. and Meier, R. J. (1987). Oxygen-induced secondary substituent effects in polymer XPS spectra. *J. Elec. Spec.*, **43**, 131–7.

Reilley, C. N., Everhart, D. S., and Ho, F. F.-L. (1982). ESCA analysis of functional groups on modified polymer surfaces. In *Applied electron spectroscopy for chemical analysis* (ed. H. Windawi and F. F.-L. Ho), pp. 105–33, John Wiley and Sons, New York.

Seah, M. P. and Hondros, E. D. (1977). Segregation to interfaces. *Int. Metal. Revs*, **22**, 262–301.

Watts, J. F. (1985). Analysis of ceramic materials by electron spectroscopy. *J. Microscopy*, **140**, 243–60.

Watts, J. F. (1987). The use of X-ray photoelectron spectroscopy for the analysis of organic coating systems. In *Surface Coatings I* (ed. A. D. Wilson, J. W. Nicholson, and H. J. Prosser), pp. 137–87. Elsevier Applied Science Publishers Ltd, London.

Watts, J. F. (1988). The application of surface analysis to studies of the environmental degradation of polymer-to-metal adhesion. *Surf. Interf. Anal.*, **12**, 497–503.

West, R. H. and Castle, J. E. (1982). The correlation of the Auger parameter with refractive index: an XPS study of silicates using ZrL$\alpha$ radiation. *Surf. Interf. Anal.*, **4**, 68–75.

The Proceedings of the Biennial European Conference on Applications of Surface and Interface Analysis (ECASIA) is published as a single bound volume of Surface and Interface Analysis, (ECASIA 85 Vol. 9, ECASIA 87 Vol. 12 and ECASIA 89 Vol. 16). These proceedings provide a timely overview of the application of surface analysis in all aspects of materials science.

## Chapter 6

Baun, W. L. (1981). Ion scattering spectrometry: a versatile technique for a variety of materials. *Surf. Interf. Anal.*, **3**, 243–50.

Briggs, D. (1986). SIMS for the study of polymer surfaces: a review. *Surf. Interf. Anal.*, **9**, 391–404.

Budd, P. M. and Goodhew, P. J. (1988). *Light-element analysis in the transmission electron microscope: WEDX and EELS*. Oxford University Press, Oxford.

Castle, J. E. and Castle, M. D. (1983). Simultaneous XRF and XPS analysis. *Surf. Interf. Anal.*, **5**, 193–8.

Chu, W. K., Mayer, J. W., and Nicolet, M.-A. (1978). *Backscattering spectrometry*. Academic Press Inc., New York.

Clarke, N. S., Ruckman, J. C., and Davey, A. R. (1986). The application of laser ionization mass spectrometry to the study of thin films and near-surface layers. *Surf. Interf. Anal.*, **9**, 31–40.

Degreve, F., Thorne, N. A., and Lang, J. M. (1988). Metallurgical applications of SIMS. *J. Mater. Sci.*, **23**, 4181–208.

Goodhew, P. J. and Castle, J. E. (1983). A survey of physical examination and analysis techniques. Inst. Phys. Conf. Ser. No. 68 (EMAG), pp. 515–22.

Goodhew, P. J. and Humphreys, F. J. (1988). *Electron microscopy and analysis.* Taylor and Francis Ltd, London.

Vickerman, J. C. (1987). Secondary ion mass spectrometry. *Chemistry in Britain*, (10), 969–74.

Walls, J. M. (1989). *Methods of surface analysis.* Cambridge University Press, Cambridge.

Werner, H. W. and Garten, R. P. H. (1984). A comparative study of methods for thin-film and surface analysis. *Rep. Prog. Phys.*, **47**, 221–344.

# Index

# Appendices

## Appendix 1.    Chart of principle Auger electron energies

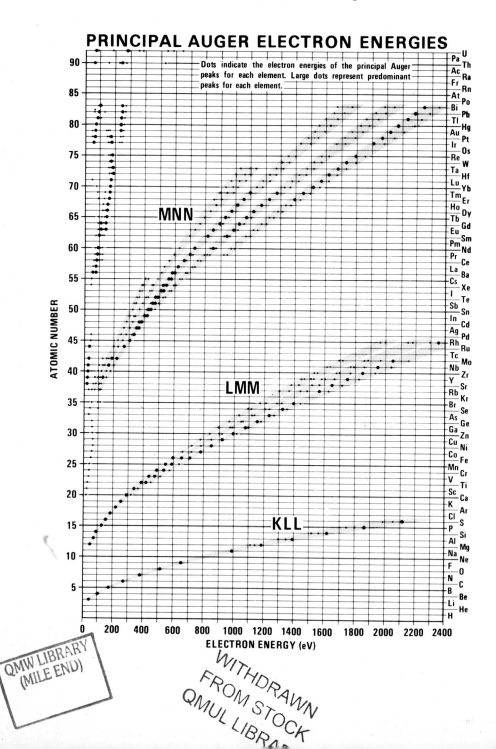